SpringerBriefs in Computer Science

SpringerBriefs present concise summaries of cutting-edge research and practical applications across a wide spectrum of fields. Featuring compact volumes of 50 to 125 pages, the series covers a range of content from professional to academic.

Typical topics might include:

- A timely report of state-of-the art analytical techniques
- A bridge between new research results, as published in journal articles, and a contextual literature review
- A snapshot of a hot or emerging topic
- An in-depth case study or clinical example
- A presentation of core concepts that students must understand in order to make independent contributions

Briefs allow authors to present their ideas and readers to absorb them with minimal time investment. Briefs will be published as part of Springer's eBook collection, with millions of users worldwide. In addition, Briefs will be available for individual print and electronic purchase. Briefs are characterized by fast, global electronic dissemination, standard publishing contracts, easy-to-use manuscript preparation and formatting guidelines, and expedited production schedules. We aim for publication 8–12 weeks after acceptance. Both solicited and unsolicited manuscripts are considered for publication in this series.

**Indexing: This series is indexed in Scopus, Ei-Compendex, and zbMATH **

Marwan Omar

Machine Learning for Cybersecurity

Innovative Deep Learning Solutions

 Springer

Marwan Omar
Department of ITM and Cybersecurity
Illinois Institute of Technology
Chicago, IL, USA

ISSN 2191-5768   ISSN 2191-5776   (electronic)
SpringerBriefs in Computer Science
ISBN 978-3-031-15892-6   ISBN 978-3-031-15893-3   (eBook)
https://doi.org/10.1007/978-3-031-15893-3

This Springer imprint is published by the registered company Springer Nature Switzerland AG
The registered company address is: Gewerbestrasse 11, 6330 Cham, Switzerland

First and foremost, I would like to dedicate this book to my wonderful wife, Maha, for inspiring me to pursue a book project and for supporting me throughout this journey. Without her unwavering support, this project would not have seen the light!
Second, this book is also dedicated to my amazing kids: Tala and Adam for always inspiring me to pursue the impossible.
Third, I dedicate this work to my caring and lovely parents: Gozi and Dahar for always encouraging me to be my best and to pursue my educational dreams.
Finally, this book is also dedicated to my lovely brothers: Faysal, Mazin, Maher, Hazim, and Sinan for always believing in me and supporting me in my educational journeys.

Contents

**1 Application of Machine Learning (ML) to Address
Cybersecurity Threats** 1
   1.1 Introduction 1
   1.2 Methodological Approach 2
      1.2.1 Review of Literature 3
   1.3 Conclusion 10
   References. 10

**2 New Approach to Malware Detection Using Optimized
Convolutional Neural Network** 13
   2.1 Introduction 13
      2.1.1 Need for the Study. 16
      2.1.2 Major Contributions of the Study 16
   2.2 Related Work 17
   2.3 System Architecture. 20
   2.4 Methodology and Dataset 27
   2.5 Empirical Results. 29
      2.5.1 Improving the Baseline Model. 30
      2.5.2 Finalizing Our Model and Making Predictions 32
   2.6 Results Comparison with Previous Work. 33
   2.7 Conclusion 33
   References. 34

**3 Malware Anomaly Detection Using Local Outlier Factor
Technique.** 37
   3.1 Introduction 37
      3.1.1 Malware Detection. 38
      3.1.2 Intrusion Detection Systems 38
      3.1.3 Network-Based Intrusion Detection System 39
      3.1.4 Advantages of Network-Based Intrusion
            Detection System. 40

3.2   Related Works 40
3.3   Proposed Methodology 42
      3.3.1   Local Outlier Factor............................. 42
3.4   Results and Discussion 44
3.5   Conclusion 47
References... 47

Chapter 1
Application of Machine Learning (ML) to Address Cybersecurity Threats

Abstract As cybersecurity threats keep growing exponentially in scale, frequency, and impact, legacy-based threat detection systems have proven inadequate. This has prompted the use of machine learning (hereafter, ML) to help address the problem. But as organizations increasingly use intelligent cybersecurity techniques, the overall efficacy and benefit analysis of these ML-based digital security systems remain a subject of increasing scholarly inquiry. The present study seeks to expand and add to this growing body of literature by demonstrating the applications of ML-based data analysis techniques to various problem domains in cybersecurity. To achieve this objective, a rapid evidence assessment (REA) of existing scholarly literature on the subject matter is adopted. The aim is to present a snapshot of the various ways ML is being applied to help address cybersecurity threat challenges.

Keywords Machine learning security · Deep learning algorithms · AI for cybersecurity · Data analytics and cybersecurity · Cyberattacks · Security threats

1.1 Introduction

The damage—both immediate and long term—that cybersecurity threats can wreak upon individuals, organizations, and even governments can be huge and incredibly costly. For instance, in 2021, the latest year for which such information is available, studies reveal that cybersecurity attacks caused severe work interruptions or production downtime in at least 47% of the organizations affected. Loss of personally identifiable information (henceforth, PII) was another major impact and affected close to 46% of organizations (Fig. 1.1). Collectively, the financial implications of these disruptions are huge, and this has been increasing significantly over the recent past. For example, between 2015 and 2020, cybercrime-related damages reported to the Internet Crime Complaint Center (IC3) rose from an estimated $1 billion to over $4.2 billion [9]. Besides individual and organizational impacts, cybersecurity breaches also threaten the very foundation of financial infrastructures and could

M. Omar, *Machine Learning for Cybersecurity*, SpringerBriefs in Computer Science, https://doi.org/10.1007/978-3-031-15893-3_1

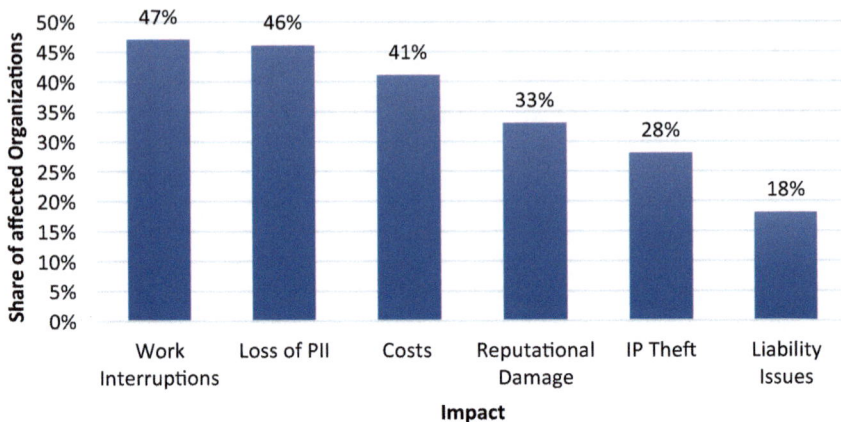

Fig. 1.1 Cybersecurity attack impacts on organizations worldwide. The graph depicts the major impacts of cybersecurity attacks on organizations worldwide in 2021. Besides work interruptions, cyberattacks can also lead to loss of customer PII (46%), additional costs for external services to address the issues (41%), theft of intellectual property (IP) (28%), sustained productivity impairment (22%), and business shutdown (15%). (Source: Sava [16])

even pose significant threats to national economies and security if deployed by adversarial agents, including state actors and terrorist organizations, and directed towards critical infrastructures like transportation and energy systems [3]. Mitigating cyber threats has therefore been one of today's major pressing concerns.

But as cybersecurity threats keep growing exponentially in scale, frequency, and impact, legacy-based threat detection systems have proven inadequate [14]. This has prompted the use of machine learning (hereafter, ML) to help address the problem [3, 7]. But as organizations increasingly use intelligent cybersecurity techniques, the overall efficacy and benefit analysis of these ML-based digital security systems remain a subject of increasing scholarly inquiry. The present study seeks to expand and add to this growing body of literature by demonstrating the applications of ML-based data analysis techniques to various problem domains in cybersecurity. To achieve this objective, a rapid evidence assessment (REA) of existing scholarly literature on the subject matter is adopted. The aim is to present a snapshot of the various ways ML is being applied to help address cybersecurity threat challenges.

1.2   Methodological Approach

Market data divulge that digital security threats, from malware and virus attacks to more sophisticated forms of cyber assaults like distributed denial-of-service (DDoS) and advanced persistent threats (APTs), keep growing exponentially in scale, frequency, and overall impact. According to research, including seminal findings by Rupp [14], one major reason for this is the growing size of the cybersphere, a phenomenon that has profoundly expanded the available threat surface (e-Commerce, IoT, telecommuting, and BYOD, among others). As such, literary everything about contemporary personal and professional life has become inherently susceptible to cybersecurity risks. In addition, cybercriminals are becoming more sophisticated, coordinated, and well-resourced, even by nation-states. Together with the "cybercrime-as-a-service" model, DevOps deployed by attackers, and cloudification of almost every computing service, these factors, along with several others such as the proliferation of cryptos, have enabled cyberattackers to not only (1) accumulate budget and data but also (2) invest in R&D to create more enhanced and impactful attack models with higher volume, diversity, and velocity [14]. The response has been the development and use of intelligent cybersecurity systems, such as ML. To identify how ML is used in this regard, an REA methodological approach is adopted.

REA is a rapid review of the available literature approach that presents evidence-based solutions or information about a particular topic of interest. This technique now proposed and advanced by the Government Social Research (GSR) website as a way of providing an evaluation of what is known about a particular issue by systematically reviewing and critically appraising existing research is finding widespread applications across multiple domains, including computer science [6]. Besides its shortcomings, such as lack of breadth in both scope and content, REA offers a rigorous and explicit method of identifying evidence across multiple pieces of literature to answer a research question (RQ) of interest [6]. In our case, the RQ is what is the role of ML application in addressing cybersecurity threats. As such, pertinent literature from e-databases such as ScienceDirect, Wiley Online Library, Google Scholar, and Elsevier were critical assessments for evidence on the potential application of ML in cybersecurity.

1.2.1 Review of Literature

In recent years, interest and overall progress in the field of ML and broader artificial intelligence (hereafter, AI) have increased significantly, with novel applications exuberantly pursued across multiple sectors [13]. At the same time, the digital communication technologies on which the world has come to depend on present numerous security concerns: cyberattacks have only escalated in number but also in frequency and scale of impact, drawing mounting attention to the susceptibilities of cyber systems and the significance and the overall need to boost their security [13]. Over the past 5 years, for instance, the IC3 has received at least 552,000 cybersecurity complaints every year (Fig. 1.2).

These complaints cover a broad range of issues affecting internet users worldwide. In the light of this rapidly evolving landscape, there is a significant and legitimate concern among policymakers, researchers, and security practitioners about the potential application of ML for enhancing cybersecurity [13]. Several studies have explored this issue.

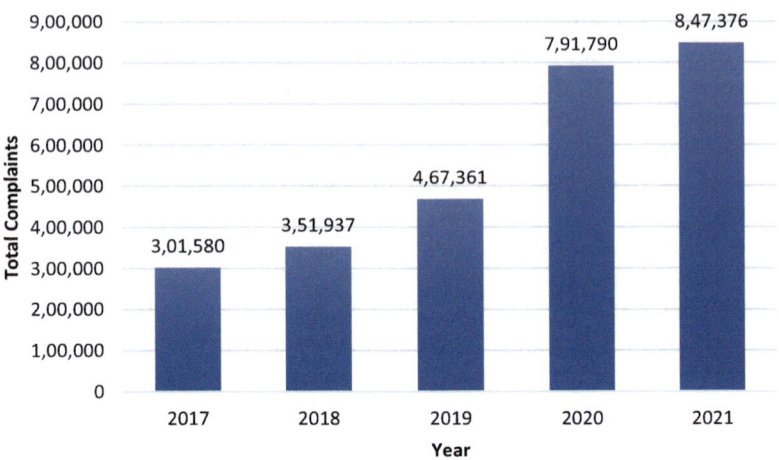

Fig. 1.2 Complaints and losses over the past 5 years. (Source: Federal Bureau of Investigation [FBI] [4])

Handa et al. [8] conducted a literature review on some of the applications of machine learning (ML) in cybersecurity. They also examined the adversarial threats to the use of machine learning. The review focuses on applications, including (1) power system security, (2) cyberattack detection on industrial control systems (ICS) by zone divisions, (3) detection of intrusion on SCADA systems, (4) detection of intrusion on VANETs (vehicular ad hoc network), and (5) malware analysis. Power system security involves protecting electric power systems from intentional attacks that target stressed power systems intending to create cascading blackouts. Here, decision-based ML algorithms can enhance the cyberattack surface and detect stressed conditions in these systems by analyzing predictors such as voltage magnitudes, current magnitudes, and angle differences. ML can also help determine the ideal number of phasor measurement units (PMUs) (devices for estimating the magnitude and phase angles of current or voltage) needed for deployment in the power system and the most critical locations where they are needed.

In detecting cyberattacks on ICSs, ML algorithms can create an automatic intrusion detection system (IDS) for securing critical infrastructure. The most vulnerable component in ICSs is the programmable logic controller (PLC)—a small computer fitted with an operating system for controlling machine operations [8]. If worms attack the PLC, they could result in compromising the working of the machines and affect operations. Here, ML-based IDS can detect concealed cyberattacks by training the algorithms to detect abnormal patterns in the data being collected by the system.

Thirdly, ML can detect ongoing intrusion on SCADA systems used for monitoring critical infrastructures such as electric grid and water transmission systems. ML-based IDS can identify attacks and initiate alerts that help in remedial action. An example is the support vector machine—an ML algorithm for distinguishing between normal and malicious traffic. ML-based IDS can also be developed for VANETs (vehicular ad hoc network)—a wireless network connecting vehicles to roadside base stations to provide valuable safety and traffic information [1]. These networks are vulnerable to attacks such as eavesdropping and interference. Here, a distributed ML-based IDS can help detect abnormal or malicious behavior on the network by allowing nodes to share their threat classification data.

Lastly, ML can be used in malware analysis, especially polymorphic and metamorphic malware that keep altering their structure or code after each infection. Of concern is the threat posed by zero-day malware—newly discovered vulnerabilities that attackers can exploit before software developers can detect and address them [10]. Such attacks cannot be addressed by signature-based approaches. Here, ML algorithms can detect known malware and provide information to help detect new malware. ML-based malware analysis works in two stages: (1) the training stage, where the algorithm extracts key information about benign and malicious data and uses it to create a predictive model, and (2) the testing stage, where the created predictive model evaluates unknown data to determine whether they pose a threat to the system (see Fig. 1.1). The common ML techniques used in detecting second-generation malware include decision tree, neural networks, deep learning, and data mining.

Ford and Siraj [5] examined machine learning applications for cybersecurity. These applications include network intrusion detection, phishing detection, profiling of smart-meter energy use, detecting spam in social networks, cryptography, human interaction proofs, and keystroke-based authentication. In phishing detection, ML-based classifiers such as logic regression, neural networks, random forests, support vector machines, and Bayesian regression trees can be used. Of these, logic regression has the highest precision rate. In network intrusion detection, ML-based detection systems are preferable because they can adapt to new and unknown attacks. For instance, such systems may combine support vector machines (SVMs) with misuse detection approaches to enhance the identification of anomalies.

Mathew [11] investigated the state of ML use in addressing the probability of cyber threat actualization. Through a literature review, the study identified four main areas of ML application in cybersecurity: (1) spam detection, (2) malware analysis, (3) intrusion detection system (IDS), and (4) android (mobile) malware detection.

Das and Morris [2] examined how machine learning can be used in cyber analytics to detect intrusion, classify traffic, and email filtering. The findings identified the common ML techniques for cybersecurity, including Bayesian network, decision trees, clustering, artificial neural networks (ANN), genetic algorithm and programming, inductive learning, and the hidden Markov model (HMM). For example, a Bayesian network can be applied to detecting anomalies and known attack patterns and signatures. Preliminary tests on some sample Bayesian network models have revealed an 89% performance rate with 99% and 89% detection rate for probe and denial-of-service (DoS) attacks. ML decision trees can be used to increase traffic processing speeds by allowing parallel evaluation and comparisons with attack signatures. Clustering algorithms can be used in the real-time detection of known attack signatures. Sample models of such algorithms have also demonstrated a 70–80% accuracy in detecting unknown (zero-day) attacks [2]. Overall, the study recommended machine learning algorithms for three cybersecurity domains: (1) misuse detection (e.g., decision trees and genetic algorithms), (2) anomaly detection (i.e., clustering algorithm), and (3) intrusion detection systems.

Salloum et al. [15] reviewed literature surveys on the machine learning and deep learning techniques that can be used in detecting network intrusion. Like Das and Morris [2], the study identified decision trees, logic regression, naïve Bayes, and random forest among the common ML algorithms used to detect network intrusion. Deep learning—a new ML branch—is more effective in detecting flow-based anomalies. The common cybersecurity datasets used with ML include the following: (1) the ISOT (information security & object technology) dataset that comprises 1,675,424 total traffic flow and can detect abnormal traffic; (2) the HTTP CSIC 2010 dataset for web attack detection and comprises about 6000 normal and 25,000 anomalous requests; (3) the CTU-13 dataset comprising 13 different malware signatures for detecting botnet traffic; and (4) the UNSW-NB15 dataset comprising an hour-long anonymized traces for a DDoS attack and 9 major attacks such as worms, reconnaissance, DoS, backdoors, and fuzzers. Thus, ML and its unique datasets can enhance cybersecurity in modern networks.

Lastly, Musser and Garriott [12] assessed the potential of machine learning to detect and intercept cybersecurity attacks at higher rates than traditional approaches. Using a four-stage cybersecurity model—prevention, detection, response and recovery, and active defense—the study analyzed the impact of recent ML innovations. Four findings emerged from the analysis. First, ML is more accurate in detecting and triaging potential attacks. Second, ML can enable the partial or full automation of several cybersecurity tasks, such as vulnerability discovery and attack disruption. Third, ML can potentially offer incremental cybersecurity advances from previously underutilized defensive strategies. Fourth, ML could, unfortunately, change the threat landscape and make certain strategies more attractive to attackers.

Traditionally, ML has been used for three tasks—spam, intrusion, and malware detection [12]. ML-based spam detection provides better spam classifiers by extracting more technical words or phrases from mail headers (e.g., server information and IP addresses). More complex algorithms can track the past email transactions of users and flag anomalous contacts or assess the authenticity of branded emails using deep learning models. In intrusion detection, ML can be used to automate certain types of misuse-based detection (identifying attacks based on resemblance to previous attacks) by learning the telltale signs of different kinds of attacks, thus eliminating the need for humans to create the list of rules that would trigger an alert. In contrast, anomaly-based detection flags any behavior that differs from normal or baseline operations. In malware detection, ML can help in scanning and matching file contents with a list of malware signatures. Newer ML architectures such as deep learning, generative adversarial networks (GANs), reinforcement learning, and massive natural language models are being adopted across the four stages of cybersecurity (see Figs. 1.3, 1.4, 1.5, and 1.6).

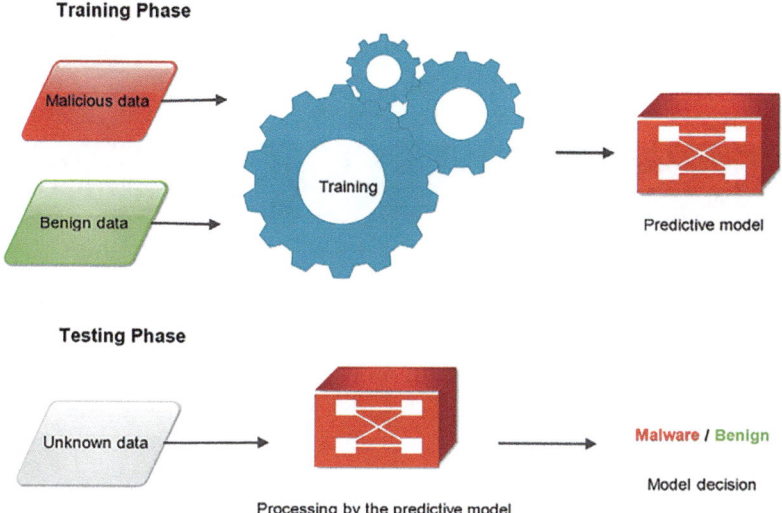

Fig. 1.3 ML-based malware analysis. (Source: Handa et al. [8])

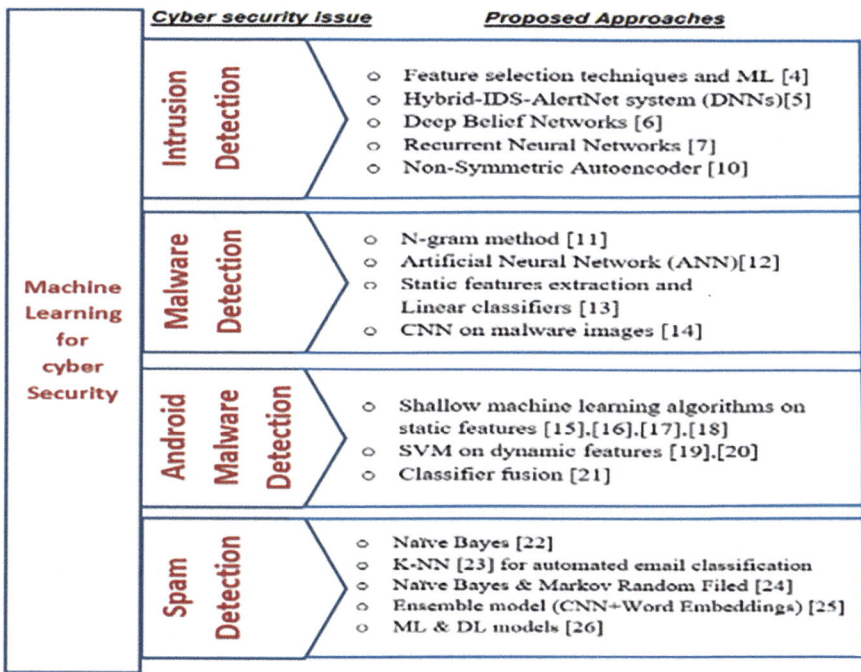

Fig. 1.4 The main ML algorithms for addressing specific cybersecurity issues. (Source: Mathew [11])

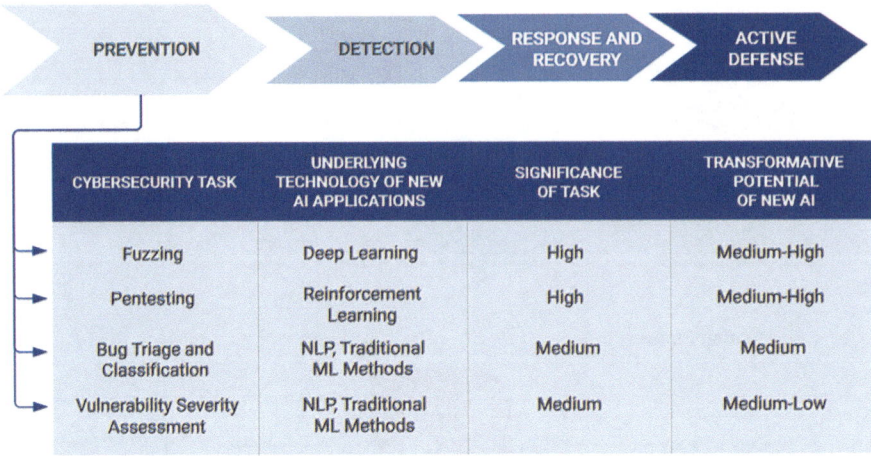

Fig. 1.5 New ML applications for prevention. (Source: Musser and Garriott [12])

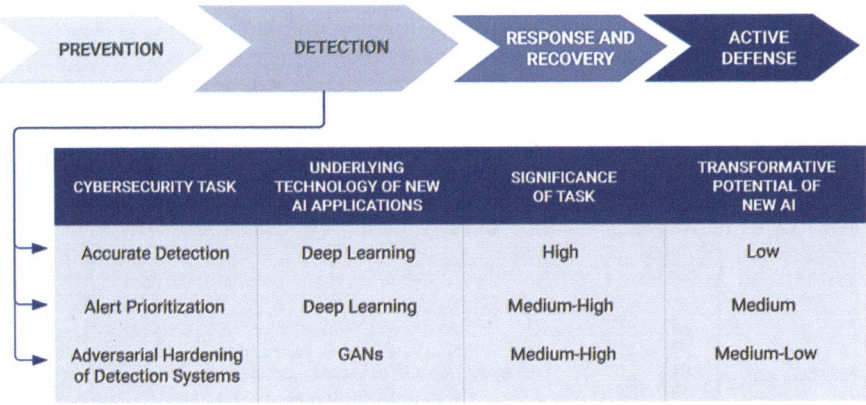

Fig. 1.6 New ML applications for detection. (Source: Musser and Garriott [12])

Prevention involves searching for and patching vulnerabilities. For instance, fuzzers search for vulnerabilities in codes while pentesting (penetration testing) searches for publicly known vulnerabilities as well as insecure configurations within networks. For large organizations that run on multiple codes, pentesting can be costly and time-consuming, hence the use of AI agents based on reinforcement learning to conduct such tests more strategically. Bug triage and severity assessment involve the use of ML in identifying the most critical vulnerabilities.

Detection is the phase where deep learning and newer ML models could have a transformative impact. However, such profound breakthroughs have not occurred yet as many firms still use simpler detection models. The minimal usage of deep learning today involves providing sophisticated analysis that can help in better prioritization of threats. Accurate analyses mean fewer false alerts for investigation. In

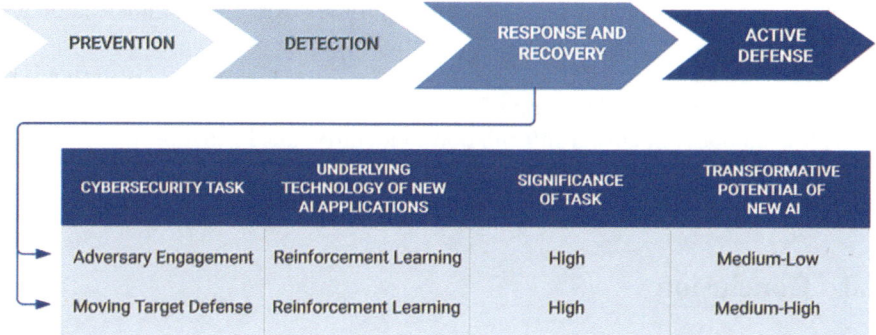

Fig. 1.7 New ML applications for response and recovery. (Source: Musser and Garriott [12])

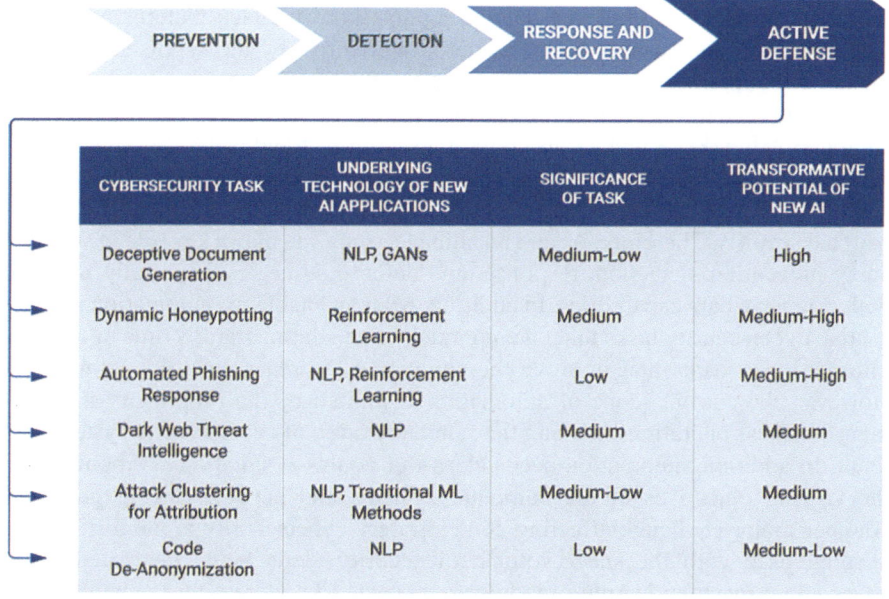

Fig. 1.8 New ML applications for active defense. (Source: Musser and Garriott [12])

addition, GANs can be used to harden ML systems to anticipate potential attacks (Fig. 1.7).

AI and ML can support response and recovery in two ways: (1) accurate categorization of ongoing attacks and choosing the suitable response strategy and (2) automation of decisions such as imposing user restrictions to limit infections and isolating a compromised machine from a network (Fig. 1.8).

Active defense involves strategies used in responding to new threats. Three areas of defense apply to ML: (1) deception, (2) threat intelligence, and (3) attribution [12]. Deception is the simplest defense mechanism involving the generation of realistic-looking documents, files, or activity profiles using ML to mislead attackers. Threat intelligence deals with gathering information on potential adversaries to anticipate attacks and create stronger defenses. ML can assist by clustering dark web users or text mining. Attribution involves tracing attacks to specific adversaries. Here, ML models can cluster information to identify attacks that bear similar attributes or descriptions of a previous attack.

1.3 Conclusion

The pervasive use of internet-connected devices, systems, and networks has intensified research on cybersecurity strategies. While traditional cybersecurity methods such as antivirus/anti-malware programs, encryption, firewalls, and intrusion detection systems still offer baseline defense against known attack patterns and signatures, they are increasingly being rendered ineffective by the sophisticated cyber threats that continue to emerge today. As information and communication technologies advance, so do the methodologies and techniques used by adversaries to attack computer networks. A dangerous consequence of this trend is the magnified risk of zero-day attacks, which pose a grave threat to internet-connected industrial control systems and critical infrastructure that, in turn, threaten societal stability and national security. Therefore, AI and machine learning can play a key role in strengthening prevention, detection, response, and defense strategies, especially for firms with extensive data capabilities. In addition, ML can enable the automation of many routine cybersecurity tasks that take up much of an administrator's time or add new information streams that improve the situational awareness of the administrator. However, the current scope of automation is limited by the proprietary nature of many business operations, making the standardization of cybersecurity systems difficult. In addition, automation needs clarity of goals—a situation compounded by the varying objectives of the computer systems and networks being protected. Another major challenge affecting contemporary cybersecurity is the difficulty in keeping pace with the sheer volume of security alerts being generated. More advances in machine learning models are expected to address these challenges in the future, even though the innovations are likely to be incremental, as proven by past developments. These advances will also undoubtedly enhance the training of new cybersecurity experts.

References

1. Chadha, R. D. (2015). Vehicular ad hoc network (VANETs): A review. *International Journal of Innovative Research in Computer and Communication Engineering, 3*(3), 2339–2346. https://www.rroij.com/open-access/vehicular-ad-hoc-network-vanets-a-review.pdf

2. Das, R., & Morris, T. H. (2017). Machine learning and cyber security. *2017 International Conference on Computer, Electrical & Communication Engineering (ICCECE)*. https://doi.org/10.1109/iccece.2017.8526232

3. Dua, S., & Du, X. (2016). *Data mining and machine learning in cybersecurity*. CRC Press.

4. Federal Bureau of Investigation (FBI). (2021). *Internet Crime Report 2021*. https://www.ic3.gov/Media/PDF/AnnualReport/2021_IC3Report.pdf

5. Ford, V., & Siraj, A. (2014). Applications of machine learning in cyber security. *ISCA 27th International Conference on Computer Applications in Industry and Engineering (CAINE-2014)*. https://vford.me/papers/Ford%20Siraj%20Machine%20Learning%20in%20Cyber%20Security%20final%20manuscript.pdf

6. Grant, M. J., & Booth, A. (2009). A typology of reviews: An analysis of 14 review types and associated methodologies. *Health Information and Libraries Journal, 26*(2), 91–108. https://doi.org/10.1111/j.1471-1842.2009.00848.x

7. Halder, S., & Ozdemir, S. (2018). *Hands-on machine learning for cybersecurity: Safeguard your system by making your machines intelligent using the Python ecosystem*. Packt Publishing.

8. Handa, A., Sharma, A., & Shukla, S. K. (2019). Machine learning in cybersecurity: A review. *WIREs. Data Mining and Knowledge Discovery, 9*(4). https://doi.org/10.1002/widm.1306

9. Johnson, J. (2021, March 18). *Cybercrime: Reported damage to the IC3 2020*. Statista. Retrieved 04 May 2022 from https://www.statista.com/statistics/267132/total-damage-caused-by-by-cyber-crime-in-the-us/

10. Kaspersky. (2021, June 17). *What is a zero-day attack?—Definition and explanation*. www.kaspersky.com. Retrieved 04 May 2022 from https://www.kaspersky.com/resource-center/definitions/zero-day-exploit

11. Mathew, A. (2021). Machine learning in cyber-security threats. *SSRN Electronic Journal*. https://doi.org/10.2139/ssrn.3769194

12. Musser, M., & Garriott, A. (2021). *Machine learning and cybersecurity: Hype and reality*. Center for Security and Emerging Technology (CSET). https://cset.georgetown.edu/wp-content/uploads/Machine-Learning-and-Cybersecurity.pdf

13. National Academies of Sciences (NAS). (2020). *Implications of artificial intelligence for cybersecurity: Proceedings of a workshop*. National Academies Press.

14. Rupp, M. (2022, March 4). *Machine learning and cyber security: An introduction*. VMRay. Retrieved 04 May 2022 from https://www.vmray.com/cyber-security-blog/machine-learning-and-cyber-security-an-introduction/

15. Salloum, S. A., Alshurideh, M., Elnagar, A., & Shaalan, K. (2020). Machine learning and deep learning techniques for cybersecurity: A review. *Advances in Intelligent Systems and Computing*, 50–57. https://doi.org/10.1007/978-3-030-44289-7_5

16. Sava, J. A. (2022, February 14). *Cyber security attack impact on businesses 2021*. Statista. Retrieved 04 May 2022 from https://www.statista.com/statistics/1255679/cyber-security-impact-on-businesses/

Chapter 2
New Approach to Malware Detection Using Optimized Convolutional Neural Network

Abstract Cybercrimes have become a multibillion-dollar industry in the recent years. Most cybercrimes/cyberattacks involve deploying some type of malware. Malware that viciously targets every industry, every sector, every enterprise, and even individuals has shown its capabilities to take entire business organizations offline and cause significant financial damage in billions of dollars annually. Malware authors are constantly evolving in their attack strategies and sophistication and are developing malware that is difficult to detect and can lay dormant in the background for quite some time in order to evade security controls. Given the above argument, traditional approaches to malware detection are no longer effective. As a result, deep learning models have become an emerging trend to detect and classify malware. This paper proposes a new convolutional deep learning neural network to accurately and effectively detect malware with high precision. This paper is different than most other papers in the literature in that it uses an expert data science approach by developing a convolutional neural network from scratch to establish a baseline of the performance model first, explores and implements an improvement model from the baseline model, and finally evaluates the performance of the final model. The baseline model initially achieves 98% accurate rate, but after increasing the depth of the CNN model, its accuracy reaches 99.183 which outperforms most of the CNN models in the literature. Finally, to further solidify the effectiveness of this CNN model, we use the improved model to make predictions on new malware samples within our dataset.

Keywords Convolutional neural networks · Deep learning · Malware detection · Image features · Malware visualization · Malimg dataset · Malware classification

2.1 Introduction

The use of information technology has been a blessing to modern life as it has enabled us to reach new heights in terms of how we live and work, but it has also added considerable vulnerabilities and threats to our life. An innocuous action, such as shuffling through a malicious website or opening an email attachment, can wreak havoc and disrupt the operations of modern businesses. Not routinely updating the system or unintentionally installing malicious software can completely expose a computer system to the vulnerability and risk of cyberattacks. Cybercriminal activities have skyrocketed in recent years with hackers successfully disrupting critical operations of a sector or industry by taking an entire business organization hostage with the help of malware [1].

Ransomware is a type of malware growingly being used by cyberattackers to hold the computer system of their targets hostage until the ransom demands are fulfilled. One of the first ransomware incidents took place in 1989 when the attendees of the International AIDS conference received malware-infected floppy disks and lost their access to the files. They were then instructed to pay $189 to a specific PO Box located in Panama to get their access back to the files [2]. Today the trends of ransomware attacks have grown significantly as cyberattackers are making high-value target attacks by targeting specific organizations with significant sensitive information or financial resources. These organizations are critical to the economy of a nation like the USA. If the system of these target entities is held hostage, it affects the day-to-day operations of these critical sectors, and therefore, these entities are more likely to pay ransom to restore normal operations [2]. The 2021 ransomware attack on Colonial Pipeline, the largest refined products pipeline in the USA, is an example of such high-value target attacks. Cybercrimes, thus, have become a multibillion-dollar industry in recent years. Most cybercrimes/cyberattacks involve deploying some type of malware. As antivirus technologies are becoming more robust and evolving into anti-malware software, malware creators are also coming up with more sophisticated and potent variants that are difficult to detect and can lay dormant in the background without arousing suspicion of the security controls [3].

The number of malware samples detected in the wild has been consistently growing over the past couple of years. According to research by McAfee labs, more than 1,224,628 malware threats were detected in Q4 2020, which includes a total of 7899 unique new hashes. This necessitates bolstering the fight against malware detection and prevention especially as cyber-hackers are developing new variants of malware every day [4].

Malware classification is a crucial step for determining the name, family, or type of malware, behaviors, and signatures of malware before required actions, such as removing and quarantining, can be undertaken. For malware classification, there are primarily two approaches used, including signature-based approach and behavior-based approach. Even though traditionally signature-based classification has been used more because it is precise and fast, it fails to detect malware variants produced by the application of obfuscation techniques [5], such as packing, encryption, metamorphism, and polymorphism [6]. The problem associated with signature-based classification can be resolved by the application of behavior-based classification because the behaviors of all malware variants are almost similar. However, retrieving data related to malware behaviors is time-consuming because it should be collected during malware activation.

Therefore, a new approach for malware classification that has become popular in recent times is the one based on image processing [7]. It enables the classifier to detect and classify a malware by examining the image textures of the malware. Unlike the traditional signature- or behavior-based malware detection classification through static or dynamic analysis, this approach does not really depend on studying the signature and behaviors of malware [7], and therefore, it overcomes some of the weaknesses associated with traditional malware classification of approaches.

Traditionally, malware detection and classification involve malware analysis, which refers to the process of monitoring the behavior and purpose of a malicious URL or file. There are three types of malware analysis used most, including static analysis, dynamic analysis, or a hybrid between the two. Static analysis focuses on retrieving features by inspecting an application's manifest and disassembled code [8]. Dynamic analysis monitors the behavior of an application during its execution, while hybrid analysis monitors an application before its installation and during its execution. Of the three types of analyses, hybrid analysis is the most powerful as both static and dynamic analyses alone cannot detect the threats from the most sophisticated malware variants [8]. Xu et al. [8] in their study revealed that hybrid analysis can become more powerful if it is combined with deep learning technologies. The use of deep learning models improves the performance of malware detection and classification significantly with an accuracy of 95–99% [9].

Applying machine learning, particularly deep learning models, to detect malware has been around for several years, and malware visualization has become a hot research topic among cybersecurity researchers in recent years. Different traditional machine learning approaches, including K-nearest neighbors, vector machine, random forests, decision tree, and naive Bayes, have been used for the detection and classification of a known malware [2]. Malware samples that typically come in the portable executable (PE) format (.EXE files) are constructed by a sequence of bits. Each malware binary consists of a string of zeros and ones. The zeros and ones can be converted or represented as an 8-bit vectors. The 8-bit vectors are then organized into two-dimensional matrices, which form grayscale images, a two-dimensional matrix (corresponding to the height and the width of an image). Therefore, each grayscale pixel is represented by a value ranging from 0 to 255 [10].

Malware classification and detection by using image processing methodology was first proposed by Nataraj et al., who first converted malware binaries to grayscale images [2]. The premise behind converting malware into a grayscale image is to view malware from an image processing perspective. Image-based malware classification and detection aims to detect and classify the existence of a malware binary by studying the texture of the malware image, which is easily converted from the collected binary malware. This malware detection method easily lends itself to deep learning models, such as convolutional neural networks (CNNs) that are widely used for image recognition. The beauty of this approach is that, unlike other malware detection approaches, it can even detect any small changes to malware code and, better yet, it can even detect packed as well as obfuscated malware. This is possible because when malware authors make changes to their code or pack their binaries or even obfuscate it, the texture will occur at a different position in the image representing malware [11]. When we convert malware samples into images, we can then apply deep learning algorithms to find visual patterns or similarities among malware families because malware authors often reuse code to create new variants. It's a known fact that malware is no longer written but rather assembled [12].

2.1.1   Need for the Study

As mentioned earlier, machine learning (ML), particularly deep learning models, has been extensively applied to address many of the cybersecurity challenges, including malware detection. Many deep learning algorithms have been proposed over the past few years to address malware classification and detection. Such deep learning models rely on extracting important features in a process called "feature engineering." In essence, feature engineering allows researchers to select various features from both static and dynamic analysis of malware. Features corresponding to a particular class of malware are used to train a deep learning model in order to create a separating plane between malware and clean ware [13]. Although previous research works [1, 9, 11-14] have made significant advances towards more efficient malware detection techniques using a variety of deep learning algorithms, the main issues are that most of such works develop an algorithm, apply it to a few datasets, achieve an acceptable level of accuracy, but do not strive to optimize their deep learning models. This is primarily because most cybersecurity researchers are not data scientists by trade, and therefore, they lack the expertise to apply best practices to make their learning models optimized. We strongly believe that this creates a gap in the literature where there is a lack of deep learning models with optimized characteristics. We believe that even if we develop a new learning algorithm, we should take the model to a new level and optimize it to make it more efficient and achieve higher accuracy-level results. The purpose of this research study is to bridge this gap in the literature by proposing a new convolutional deep learning neural network to detect malware accurately and effectively with high precision. This paper is different than most other papers in the literature in that it uses an expert data science approach by developing a convolutional neural network from scratch to establish a baseline of the performance model first, explores and implements an improvement model from the baseline model, and finally evaluates the performance of the final model.

2.1.2   Major Contributions of the Study

To fill the gap in literature, in this paper, a new convolutional deep learning network architecture for malware detection is proposed. The new CNN is based on an expert data science approach that develops a CNN from scratch and then uses data science best practices and approaches to optimize the model and ultimately achieve superior detection accuracy. Overall, the major contributions of our research work include the following:

1. We propose a new convolutional deep learning neural network to accurately and effectively detect malware with high precision.
2. We use an expert data science approach by developing a CNN from scratch to establish a baseline of the performance model first and explore and implement an improvement model from the baseline model.
3. We evaluate the performance of the final model. The baseline model initially achieves 98% accuracy rate, but after increasing the depth of the CNN model, its accuracy reaches 99.183%.
4. Our novel deep neural network (DNN) model outperforms most of the CNN models in the literature.
5. Finally, to further solidify the effectiveness and accuracy of this CNN model, we use the improved model to make predictions on new malware samples within our dataset.

The rest of the paper is organized as follows. Section 2.2 reviews the related work in malware classification models considering various approaches traditionally employed for malware detection. In Sect. 2.3, deep learning architecture is introduced to get insights into this research background. Section 2.3 presents the implementation architecture for deep learning in this research study and the statistical measures used to evaluate the performance of the classifier. Section 2.4 describes the methodology and dataset. Section 2.5 discusses the experimental study and the results obtained for malware classification using our proposed CNN model. Section 2.6 presents the comparison of our results with previous work.

Finally, Sect. 2.7 provides the conclusion of this study and future work.

2.2   Related Work

To highlight the importance of our work, we investigated the application of machine learning and deep learning techniques for the detection and classification of malware by other researchers. Nataraj et al. [1] are considered the pioneers in using machine learning for malware detection. They used a machine learning technique, k-NN, for their image-based malware classification. They used GIST descriptor to extract features from the input images. They used the much popular, but relatively small, Malimg dataset with 9339 malware binaries from 25 different malware families. In their proposed model, they achieved an impressive 98% accuracy rate. One drawback for their novel approach was that the GIST descriptor is seen as time-consuming and overly complex.

In another relevant study, Yajamanam et al. [15] took a slightly different approach by using the feature-engineering technique as an important factor to influence the accuracy of their malware classifier. So, they picked 60 features out of the 320 features available from GIST for training. Unfortunately, their model achieved only

92% accuracy because reducing features could have had a negative impact on the effectiveness and accuracy of their technique.

The authors in [16] used a different descriptor to extract features needed for pre-training their deep learning model of image recognition. They used ImageNet as their dataset which contained a whopping 1.2 million images spanning 1000 classes of malware binaries. Their thought was that the bigger the dataset is, the more effective and accurate the model would be. This was based on the premise that any given deep learning model is only as good as the training data. Much to their surprise, the proposed deep learning model achieved only 92% accuracy rate, which is not too impressive given the gigantic dataset they used and compared to other relevant models in the literature.

A lightweight approach was undertaken by Jiawei et al. [17] for the detection of distributed denial-of-service malware. The researchers converted malware binaries into grayscale images and then fed them into fine-tuned CNNs. The machine learning classifier was used in local devices for further classification of the malware binaries. However, since the signature matching system contains details of each malware sample, its database is huge and, therefore, is not efficient for IoT devices with limited resources. Once trained, only a small set of training data was required for the malware classification through machine learning, and therefore, a small two-layer shallow CNN was used by the researchers for the malware detection. The researchers achieved an average accuracy rate of 94% on the classification of benign and malicious malware.

Unlike other previous studies from the literature, where image descriptors were used, Quan Le et al. [18] did not use descriptors; instead, they trained their deep learning model using input images. In their study, they converted raw input images into one-dimensional fixed size vectors before feeding into their CNN model. Although they achieved an impressive 98% accuracy rate with their proposed malware classifier, it must be noted that converting images to one-dimensional, fixed size vectors can negatively impact the quality of images, possibly leading to information loss.

In a similar vein, the author of the paper [19] used raw malware images to train a CNN for malware classification. However, their model was flawed because they manipulated the data and balanced it in a way that provides a higher accuracy rate.

Bensaud et al. [2] used six deep learning models for the detection and classification of malware images and then compared the performance of these models with one another. The six models that the researchers ran on the Malimg dataset included Inception V3, VGG16-Net, ResNet50, CNN-SVM, MLP-SVM, and GRU-SVM. The researchers were unable to use grayscale images with the two models VGG16-Net and ResNet50, because the input layers of these images require 3, 224, 224 shapes, representing red, green, and blue (RGB) channels of the image, whereas the input layers in the grayscale images require 1, 224, 224 shapes. The figure below demonstrates the prediction accuracy of the six models:

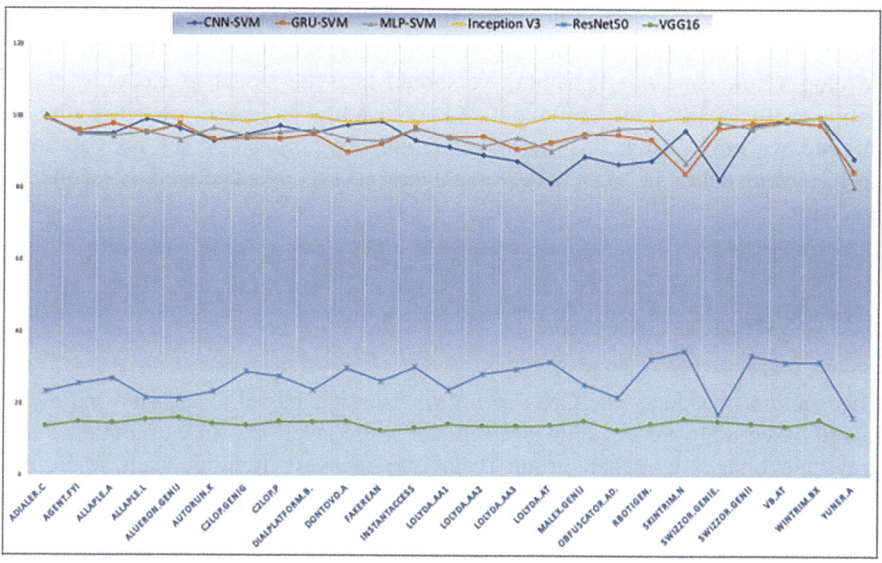

As can be seen, VGG16-Net and ResNet50 exhibited low performance in comparison with other models as these two models involved architectures that were designed to recognize colored images in RGB format. Therefore, these two models performed with the lowest accuracies when they were tested on the grayscale images. The highest accuracy of 99.24% was achieved by the Inception V3 model. CNN-SVM, GRU-SVM, and MLP-SVM also performed well at 93.22%, 94.17%, and 94.55% accuracy rates, respectively. But the accuracy rates of ResNet 50 and VGG16 were at 26.66% and 14.31%, respectively.

Lad and Adamuthe [20] proposed a combination of CNN and hybrid CNN + SVM model for malware image detection and classification. The researchers did not use softmax as an activation function. They used SVM to perform the task of malware classification based on the features retrieved by the CNN model. The proposed model generated a vector of 256 neurons with the fully connected layer, which is input to SVM. The proposed model achieved 98.03% accuracy, which is better than the accuracy rates of other CNN models, such as Xception (97.56%), InceptionV3 (97.22%), and VGG16 (96.96%) [20].

It is becoming clear from the above-stated studies that it is not necessary to use image descriptors (e.g., GIST or SIFT [21]) to extract features for image processing; rather, we can simply utilize grayscale pixel values from malware images and use those values as features for training a machine learning model. The idea here is to reduce unnecessary complexity as well as time required to build an efficient deep learning malware classification model. It must be noted, though, that for this idea to work best, we need to ensure that the dimensions of a malware input image should be around 32*32 so that we do not tax the malware classifier.

Inspired by the aforementioned challenge, this paper investigates the impact of the dimension of the input malware images and the impact of the learning technique, i.e., CNN, applied on the performance of image-based malware classifi-cation. The experiment is conducted on Malimg dataset. We believe that even if we develop a new learning algorithm, we should take the model to a new level and optimize it to make it more efficient and achieve higher accuracy-level results. To this end, we aim to bridge this gap in the literature by proposing a new convolutional deep learning neural network to detect malware accurately and effectively with high precision.

2.3   System Architecture

This section introduces the CNN as a deep learning model to classify and detect image-based malware. As shown in Fig. 2.1, we convert malware executables into grayscale images and then group them into malware families such as Botnet, Banking Trojans, Backdoor, Worm, and so on. Once this is done, we then proceed into feeding the image-based malware classifier, our CNN model.

The PE formats of the malware binaries are programs with name extensions, including .bin, .exe, and .dll [2]. PE files are identified based on their components, and these components are called .tex, .rdata, .data, and .rsrc. .tex is the first compo-nent that represents the code section and contains the instructions of the program. .rdata is the section that contains read-only data, whereas .data contains modifiable data. .rsrc is the final component of PE file standing for resources used by the mal-ware [2]. One of the simple methods for converting a binary file into a grayscale image is to treat the sequence bytes (8-bit group) of the malware binary as the pixel values of a grayscale image (encoded by 8 bits) [main paper reference]. The figure below demonstrates the sections of a malware binary in a grayscale image com-posed of textural patterns [2]. Based on the textual patterns, malware can be classified.

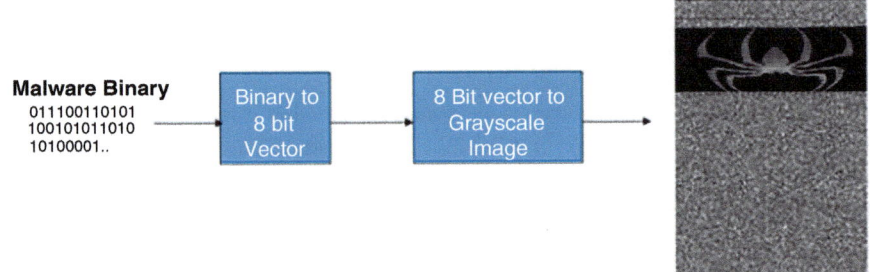

Fig. 2.1 Converting malware into an image

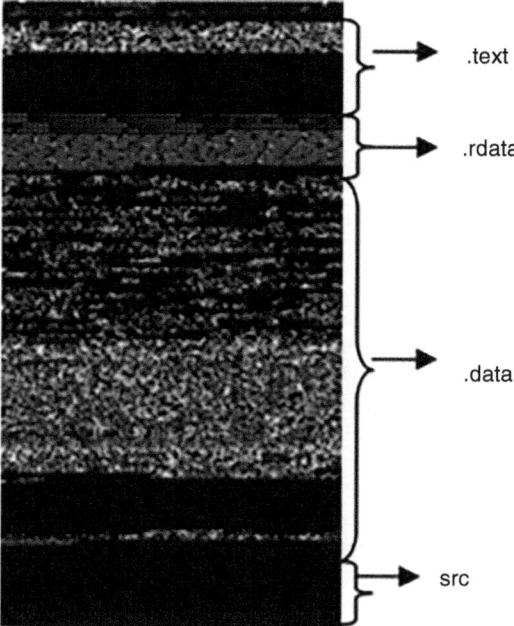

Visualization of malware binaries becomes better with images because the patterns hidden within such images become more pronounced and visible. As mentioned earlier, each grayscale pixel is represented by an 8-bit vector, the value of which can range from 00000000 (0) to 11111111 (255). Each 8-bit vector is represented by a number and can be converted into pixels in a malware image as shown below:

Although the height of a malware-based image can vary depending on the size of the malware executable, its width is typically a fixed size of 32, 64, and 128 pixels. As the width of an image is usually fixed at 32, 64, and 128 pixels [1, 6, 19].

As a result, different malware binaries generate different malware images which have different shapes as illustrated in Fig. 2.2a–c for three malware families of Alueron.gen!J, Dialplatform.B, and Swizzor.gen!E respectively.

The advantage of using image-based malware classification models is that variants of the same malware family share very similar texture as illustrated in Fig. 2.3. Three variants of the Dontovo A family look very similar to the original malware. Those three variants were randomly selected among 431 variants stored in the Malimg dataset [6]. This similarity in texture will enable the CNN model to efficiently classify malware to their respective malware family based on the similarities in image texture.

As previously highlighted, the machine learning models can be fed with grayscale malware images for training. To be precise, grayscale pixel values can be used as the features of the input images instead of extracted features from the use of image descriptors (Fig. 2.4).

Fig. 2.2 Malware image of (**a**) Alueron.gen!J; (**b**) Dialplatform.B; and (**c**) Swizzor.gen!E

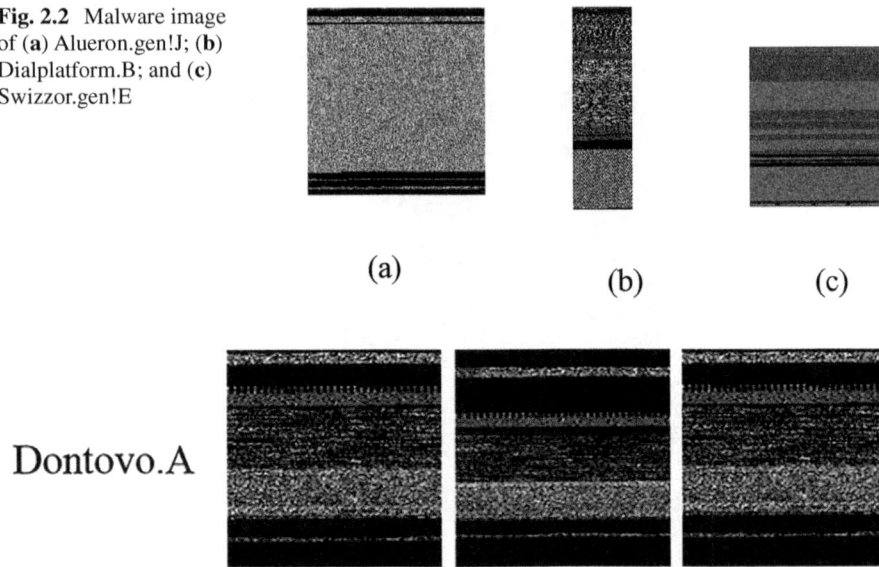

(a)   (b)   (c)

Dontovo.A

Fig. 2.3 Three variants of the Dontovo family

Fig. 2.4 The images of Dialplatform malware after normalization

Original Dialplatform

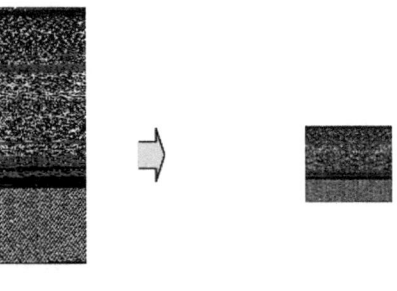

32x32   64x64

CNNs refer to a subcategory of neural networks that can effectively classify and recognize specific features from images, and therefore, CNNs are used extensively for visual image analysis. The application of CNN may range from recognition of images or videos, image classification, natural language processing [14], medical image analysis, and computer vision [9]. CNN has two primary functions, including feature extraction from images and classification of images [22]. The figure below illustrates the two functions of the CNN architecture:

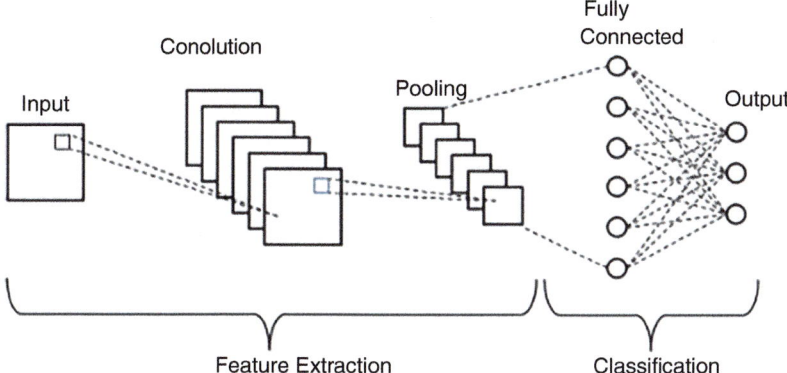

Two Functions of CNN [22]

The architecture of CNN primarily consists of two blocks [23]. The first block functions as a feature extractor by matching templates with the help of convolution filtering operations. The first layer is responsible for filtering images with many convolution kernels and produces feature maps, which are then resized or normalized. This process of filtering images and producing feature maps that are then normalized and resized is repeated several times [24]. The values derived from the last feature maps are finally used for concatenation into a vector. The output of the first block and the input of the second block are defined by this vector only. The figure below [24] demonstrates the first block, which is encircled in black:

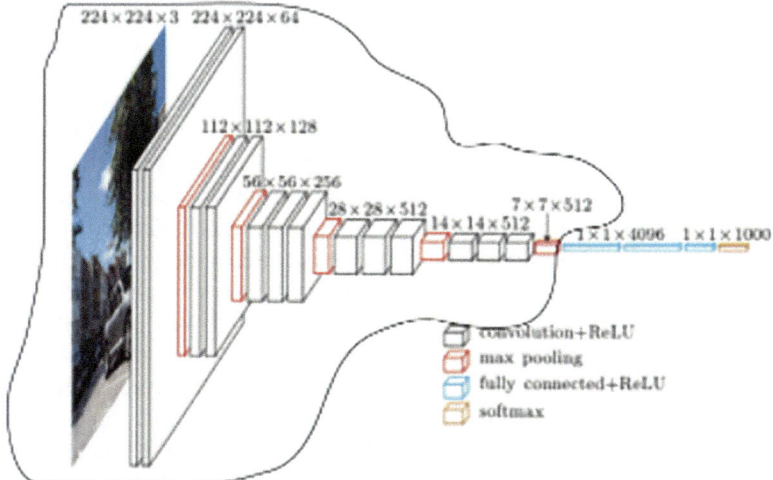

The function of the second block comes into the picture after all the neural networks are used for classification. The input factor values of the second block are transformed via several activation functions and linear combinations to generate a new vector to the output. The last vector has the same number of elements as classes. For example, element i represents the probability of the image belonging to class i.

Each element has a value ranging between 0 and 1, and the sum of all elements amounts to 1 [24]. The calculation of these probabilities is done by the last layer of the second block, which uses binary classification through a logistic function and multiclass classification through a softmax function as an activation function. As is the case with ordinary neural networks, gradient backpropagation determines the parameters of the layers. For example, during the training phase, the cross-entropy is minimized, but these parameters, in the case of CNN, refer specifically to the image features [24]. The figure below demonstrates the second block encircled in black:

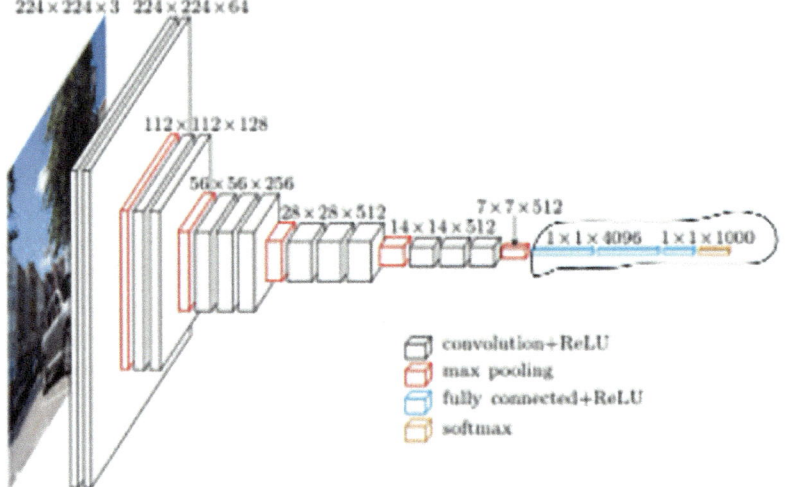

A CNN has four types of layers, including the convolutional layer, the pooling layer, the fully connected layer [25], and the ReLU correction layer [24].

The convolution layer is the first layer of CNN. The purpose of this layer is to detect the presence of a number of features in the images that are received as input [26]. The layer executes this purpose by convolution filtering. The principle behind the convolutional layer is to drag a window that represents the feature on an image and then estimate the convolution product lying between each segment of the scanned image and the feature itself. A feature is then viewed as a filter. Several images are received as input by the convolutional layer, and the convolution of each of these images is then calculated against each filter. The filters exactly represent the features we wish to see in the images. A feature map is obtained for each pair of image and filter, indicating the position of the features within the images [24]. The figure below demonstrates the convolution in which the central element of the kernel is positioned over the source pixel, which is then replaced with a weighted sum of itself and nearby pixels:

CONVOLUTION

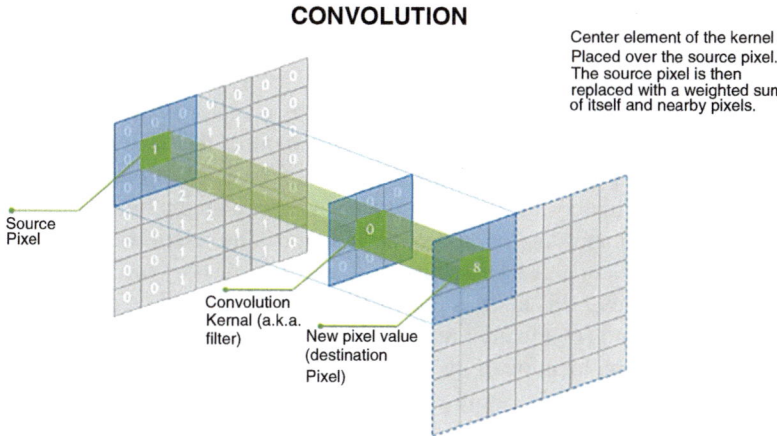

Center element of the kernel is
Placed over the source pixel.
The source pixel is then
replaced with a weighted sum
of itself and nearby pixels.

Source
Pixel

Convolution
Kernal (a.k.a.
filter)

New pixel value
(destination
Pixel)

The positioning of the pooling layer is between two convolutional layers. The pooling layer receives many feature maps and then executes the pooling operations for each of them. The pooling operation involves minimizing the image sizes without affecting their important characteristics. The pooling layer achieves this by cutting the image into cells of regular sizes and maintaining the maximum value within each cell. The most common sizes of cells are 2×2 or 3×3 cells [24]. These cells remain separated from one another by a step of 2 pixels. By reducing the number of parameters and calculations done within the network, the pooling layer enhances the network efficiency by avoiding overlearning.

The ReLU correction layer stands for rectified linear units that refer to the real nonlinear function achieved by the application of the formula of $ReLU(x) = \max(0, x)$ [24]. The visual representation of the same is as below:

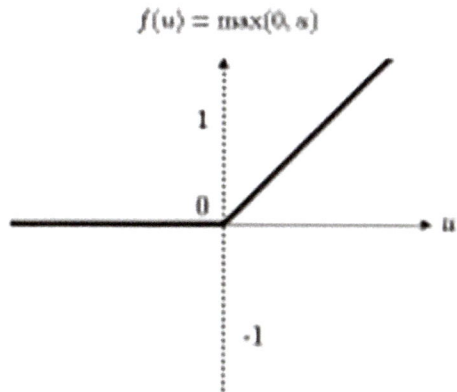

All negative values that are received as inputs by zeros are replaced by the ReLU correction layer. This layer performs the activation function.

The last layer of a CNN is the fully connected layer, which receives an input vector and generates a new output vector by applying a linear combination and an activation function for the input values received. This layer does the image classification as an input to the network and then generates a vector of size N, in which N represents the number of image classes. Each element of the vector represents the probability for the input image belonging to a class. The calculation of the probabilities is achieved by this layer by multiplying each input element by weight and then making the sum and applying an activation function. The relationship between the placement of features in an email and its class is determined by the fully connected layer. The input table of this layer is the output of the previous layer. Therefore, the input table corresponds to a feature map for a specific feature. The high values of an input table suggest the location of the feature within the image (Table 2.1).

So, the overall CNN architecture will be as in the following diagram (Fig. 2.5):

Table 2.1 Malimg dataset families [11]

	Family name	Variants
1	Allaple.L	1591
2	Allaple.A	2949
3	Yuner.A	800
4	Lolyda.AA 1	213
5	Lolyda.AA 2	184
6	Lolyda.AA 3	123
7	C2Lop.P	146
8	C2Lop.gen!G	200
9	Instantaccess	431
10	Swizzor.gen!I	132
11	Swizzor.gen!E	128
12	VB.AT	408
13	Fakerean	381
14	Alueron.gen!J	198
15	Malex.gen!J	136
16	Lolyda.AT	159
17	Adialer.C	125
18	Wintrim.BX	97
19	Dialplatform.B	177
20	Dontovo.A	162
21	Obfuscator.AD	142
22	Agent.FYI	116
23	Autorun.K	106
24	Rbot!gen	158
25	Skintrim.N	80
	Total	**9339**

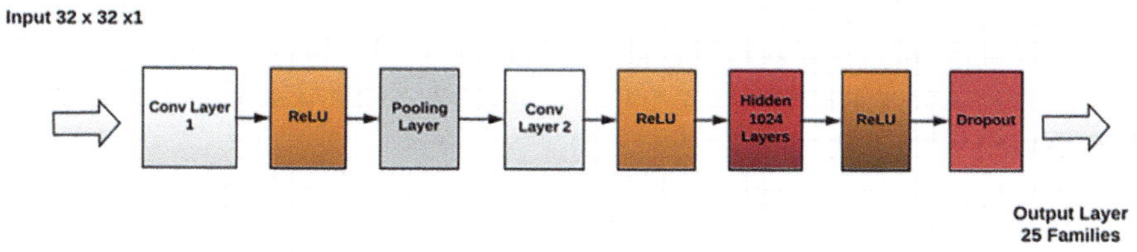
Input 32 x 32 x1
Conv Layer 1
ReLU
Pooling Layer
Conv Layer 2
ReLU
Hidden 1024 Layers
ReLU
Dropout
Output Layer
25 Families

Fig. 2.5 Overall CNN model architecture

2.4 Methodology and Dataset

The Malimg dataset has been widely used in many research projects and experiments over the past few years as it certainly lends itself well to a good deep learning convolutional neural network. Something unique about this study is that the researchers decided to develop a new CNN model from scratch instead of reviewing the literature on well-performing models. The other unique aspect here is that the researchers, unlike most studies in the literature, did not just develop a model and presented the result; instead, they go above and beyond the normal expectations by developing a CNN model from scratch, develop a robust performance evaluation of the baseline model, and explore the extensions to the baseline model in order to improve the learning capacity of the model, and finally, the researchers will develop a finalized CNN model, evaluate the finalized model, and use it to make predictions on new malware.

The Malimg dataset perfectly lends itself to our CNN model as it has a good number of train and test dataset that we can use. To gain insight into the performance of our CNN model, and the learning curve, for a given training run, we can further break up the training dataset into a train and validation dataset. We use the traditional concept of 70% of the dataset for training and the remaining 30% for testing [27].

A. Developing a Baseline Model

In order to develop a baseline CNN model for our malware classification and detection task, we will begin by developing the infrastructure for the test harness which will enable us to evaluate our model on the Malimg dataset. This step is also important as it will establish the model baseline, evaluate it, and ultimately improve the model. Developing the infrastructure for our test hardness involves the following steps: load the Malimg dataset, prepare the dataset, define our CNN model, evaluate the model, and, finally, make use of the model to make predictions using new malware samples (the holdout test) of our Malimg dataset.

B. Loading the Dataset

We know some things about the dataset. The images all have the same square size of 32 × 32 pixels, and the images are grayscale. Therefore, we can load the images and reshape the data arrays to have a single-color channel. The following code snippet illustrates how to achieve the above task:

```
# load dataset
(trainX, trainY), (testX, testY) = malmig.load_data()
# reshape dataset to have a single channel
trainX = trainX.reshape((trainX.shape[0], 32, 32, 1))
testX = testX.reshape((testX.shape[0], 32, 32 1))
```

C. Defining Our Proposed CNN Model

Now, we need to define a baseline convolutional neural network model for malware detection and classification. The model has two main aspects: the feature extraction front end comprised of convolutional and pooling layers and the classifier backend that will make a prediction to determine if a binary sample is malware or not.

We will deploy a conservative setup for the stochastic gradient descent optimizer with a learning rate of 0.01 and a momentum of 0.9. The categorical cross-entropy loss function will be optimized, suitable for multiclass classification, and we will monitor the classification accuracy metric [28].

Our model can be defined using the following code:

```
# define cnn model
def define_model():
 model = Sequential()
    model.add(Conv2D(32,    (3,    3),    activation='relu',    kernel_
initializer='he_uniform', input_shape=(28, 28, 1)))
 model.add(MaxPooling2D((2, 2)))
 model.add(Flatten())
model.
add(Dense(100,
activation='relu', kernel_initializer='he_uniform'))
 model.add(Dense(10, activation='softmax'))
 # compile model
 opt = SGD(lr=0.01, momentum=0.9)
    model.compile(optimizer=opt,    loss='categorical_crossentropy',
metrics=['accuracy'])
 return model
```

D. Evaluating Our CNN Model

Once we have completed the task of defining our CNN model for malware detection, we will now need to evaluate it. We will evaluate the model using a fivefold cross-validation. The value of $k = 5$ will ensure that we obtain a baseline for both repeated evaluation and would also ensure a relatively short running time. So, the $k = 5$ means that our training dataset will be split into five test sets [28].

Additionally, to ensure that our model will contain the same train and test datasets in each of the five folds mentioned above, we will be shuffling the training dataset before starting the split process [28]. This process will ensure that we are comparing "apples to apples" (for fair model comparison) so to speak. The batch size for training our CNN baseline model will be 32 malware samples with 10 training epochs. This setup will enable us to estimate the performance of our CNN baseline model and track the result history of each run and malware classification accuracy of each of the five folds.

The following code illustrates how to achieve the above task:

```
# evaluate a model using k-fold cross-validation
def evaluate_model(dataX, dataY, n_folds=5):
 scores, histories = list(), list()
 # prepare cross validation
 kfold = KFold(n_folds, shuffle=True, random_state=1)
 # enumerate splits
 for train_ix, test_ix in kfold.split(dataX):
 # define model
 model = define_model()
 # select rows for train and test
trainX, trainY, testX, testY = dataX[train_ix], dataY[train_ix],
dataX[test_ix], dataY[test_ix]
 # fit model
history = model.fit(trainX, trainY, epochs=10, batch_size=32, vali-
dation_data=(testX, testY), verbose=0)
 # evaluate model
 _, acc = model.evaluate(testX, testY, verbose=0)
 print('> %.3f' % (acc * 100.0))
 # stores scores
 scores.append(acc)
 histories.append(history)
 return scores, histories
```

2.5 Empirical Results

The next logical step after evaluating our model is to display and present the results. We will focus on the learning behavior of our CNN model and then estimate its performance. As it is commonly known in the deep learning community, over-fitting and under-fitting are two major issues associated with virtually any deep learning model. To this end, and to ensure that our CNN model is not flawed with neither over-fitting nor under-fitting, we have created a line plot to display and report the performance of our model on the test as well as train dataset during each of the fivefold cross-validation. The line plots will provide insight into whether or not our model is under-fitted or over-fitted for the malware dataset [19].

Only one figure with two subplots is needed to gain the insight into the over-fitting and under-fitting issue, one subplot for loss and one for accuracy of our CNN model [29]. The performance of our model on the training dataset is shown using the blue lines, while the orange lines will display our CNN model performance on the holdout test dataset. The code snippet below will achieve this task:

```
# summarize model performance
def summarize_performance(scores):
 # print summary
 print('Accuracy: mean=%.3f std=%.3f, n=%d' % (mean(scores)*100,
std(scores)*100, len(scores)))
 # box and whisker plots of results
 pyplot.boxplot(scores)
 pyplot.show()
```

We can see these cases where the model achieves perfect skill. These are good results.

```
>98.467
>98.683
>98.642
>98.850
>98.583
```

The numbers above are impressive, and they clearly show that our model evaluation is progressing.

Next, a diagnostic plot is shown, giving insight into the learning behavior of the model across each fold (Fig. 2.6).

In this case, we can see that the model generally achieves a good fit, with train and test learning curves converging. There is no obvious sign of over- or under-fitting.

Next, a summary of the model performance is calculated. We can see, in this case, the model has an estimated skill of about 98.6%, which is reasonable.

Finally, a box and whisker plot is created to summarize the distribution of accuracy scores.

2.5.1   Improving the Baseline Model

There are many ways that we might explore improvements to the baseline model. We can explore aspects of our CNN model configuration that are likely to result in an improvement, so-called low-hanging fruit. The first is a change to the learning algorithm, and the second is an increase in the depth of the model [28] (Fig. 2.7).

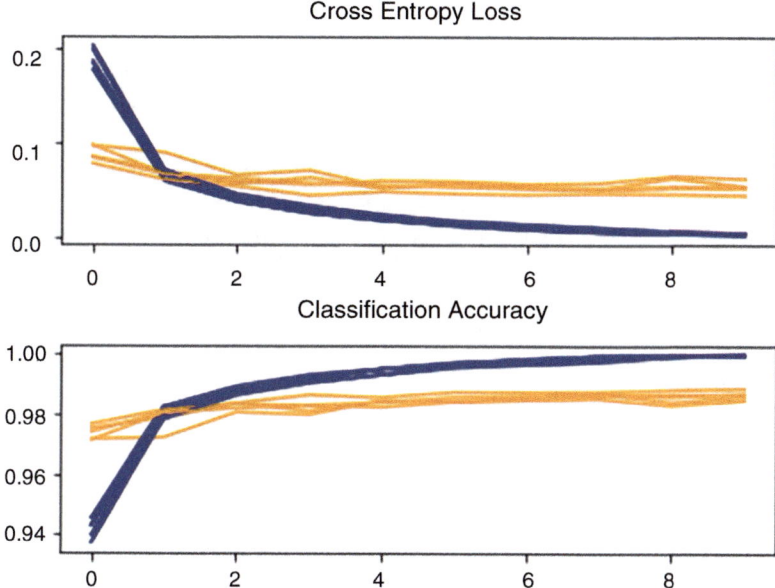

Fig. 2.6 Loss and accuracy learning curves for the baseline model during k-fold cross-validation

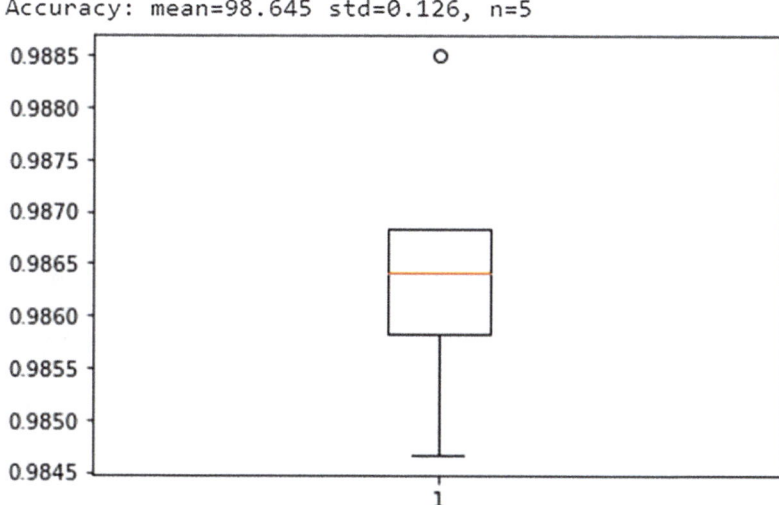

Fig. 2.7 Box and whisker plot of accuracy scores for the baseline model evaluated using k-fold cross-validation

Now we can explore a couple ways to make improvements to our CNN model. We can change the model configuration to explore improvements over the baseline model. Two common approaches involve changing the capacity of the feature extraction part of the model or changing the capacity or function of the classifier part of the model [28].

We can increase the depth of the feature extractor part of the model, following a VGG-like pattern of adding more convolutional and pooling layers with the same-sized filter, while increasing the number of filters. In this case, we will add a double convolutional layer with 64 filters each, followed by another max pooling layer [28].

Running the example reports model performance for each fold of the cross-validation process as shown below:

>98.775
>98.683
>98.967
>99.183
>99.008

The per-fold scores above may suggest some improvement over the baseline.

2.5.2   Finalizing Our Model and Making Predictions

Although it appears intriguing to continue making improvements on our model, at this point we will choose the final configuration of our CNN model. The final version of our model will be the deeper model. We have finalized our model by fitting it on the entire training Malimg training dataset and then load the model and evaluate

Table 2.2 Comparisons of the proposed DL-CNN with other existing deep learning approaches

Year	Researchers	Methods	Technique	Accuracy (%)
2011	Nataraj et al.	GIST	Machine learning	98
2017	S. Yue	CNN	Deep learning	97.32
2017	Makandar and Patrot	Gabor wavelet-kNN	Machine learning	89.11
2018	Yajamanam et al.	GIST+kNN+SVM	Machine learning	97
2018	Cui, Xue, et al.	GIST+SVM [29]	Deep learning	92.20
2018	Cui, Xue, et al.	GIST+kNN	Deep learning	91.90
2018	Cui, Xue, et al.	GLCM+SVM	Deep learning	93.20
2018	Cui, Xue, et al.	GLCM+kNN	Deep learning	92.50
2018	Cui, Xue, et al.	IDA+DRBA	Deep learning	94.50
2019	Cui, Du, et al.	CNN, NSGA-II	Deep learning	97.6
2020	Mallet	CNN, Keras	Deep learning	95.15
2020	Vasan et al.	IMCFN, color images	Deep learning	98.82
2021	Moussas and Andreatos	Image and file features, ANN	Two-level ANN	99.13
2021	Omar	CNN, Keras	Deep learning	99.18

it. We will evaluate our model performance and accuracy on the holdout test Malimg dataset. This will give us insight into how practically accurate our model performs on real malware dataset (Table 2.2).

The classification accuracy for the model on the test dataset is calculated and printed. In this case, we can see that the model achieved an accuracy of 99.180%, or just less than 1%, which is not bad at all with a standard deviation of about half a percent (e.g., 99% of scores).

>99.180

2.6 Results Comparison with Previous Work

To ensure fairness of comparison (comparing apples to apples), we provide a comparison of our deep learning CNN model's performance with other deep learning algorithms for malware classification and detection using the same dataset. The final version of our proposed convolutional deep learning model achieved an impressive accuracy score of 99.18 compared to Mallet [30] which has 95.15. This clearly shows our model is outperforming other models in the literature. It's interesting to note that Mallet [30] used almost an identical deep learning architecture that involved CNN and the Keras scientific environment. Also, our model outperforms the deep learning model proposed by Cui et al. [19] which achieved an accuracy score of 97.6. It is also interesting to note that our model even outperforms a recent work conducted by Moussas and Andreatos [31] with an impressive accuracy rate of 99.13 on the exact same dataset.

2.7 Conclusion

In this paper, we proposed, developed, and presented a deep learning convolutional neural network model for malware detection. Our model deploys a unique approach to malware detection in that it developed a deep learning CNN from scratch, developed a baseline model, developed a robust performance evaluation of the baseline model, and explored the extensions to the baseline model in order to improve the learning capacity of the model, and finally, we developed a finalized CNN model, evaluated the finalized model, and used it to make predictions on new malware.

We evaluated our model using the popular Malimg dataset of 9339 samples belonging to 25 malware families. Comparing our model's performance to existing DL-based frameworks, our results clearly illustrate that our deep learning CNN outperforms those works presented in existing deep learning-based malware detection models.

As future work, we believe that the performance of our model can be further tested and evaluated using bigger datasets such as the BIG 2015 dataset and newly made available datasets from repositories such as Kaggle. Another future direction would be to compare our model's accuracy score to the accuracy score of other similar deep learning models such as the k-NN model and others.

References

1. Ye, Y., Li, T., Adjeroh, D., & Iyengar, S. S. (2017). A survey on malware detection using data mining techniques. *ACM Computing Surveys (CSUR), 50*(3), Article No. 41.
2. Bensaoud, A., et al. (2020). *Classifying malware images with convolutional neural network models*. Department of Computer Science, University of Colorado Colorado Springs [Online]. Available https://arxiv.org/pdf/2010.16108.pdf
3. Kaspersky. (2019). *Kaspersky Security Bulletin 2019* [Online]. Available https://securelist.com/kaspersky-security-bulletin-threat-predictions-for-2019/88878/
4. Cybersecurity Ventures. (2018). [Online]. Available https://cybersecurityventures.com/-cybercrime-damages-6-trillion-by-2021/
5. Souri, A., & Hosseini, R. (2018). A state-of-the-art survey of malware detection approaches using data mining techniques. *Human-centric Computing and Information Sciences, 8*(3), 1–22.
6. Nataraj, L., Karthikeyan, S., Jacob, G., & Manjunath, B. (2011). Malware images: Visualization and automatic classification. In *Proceedings of the 8th international symposium on visualization for cyber security, Pittsburgh, Pennsylvania, USA.*
7. Lee, C., et al. (2020). An evaluation of image-based malware classification using machine learning. In *Advances in computational collective intelligence, 12th international conference, ICCCI 2020, Da Nang, Vietnam, November 30 – December 3, 2020, Proceedings.* https://doi.org/10.1007/978-3-030-63119-2_11
8. Xu, L., et al. (2016). HADM: Hybrid analysis for detection of malware. *SAI Intelligent Systems Conference 2016, 21*(22), 1037–1048.
9. Farabet, C., Couprie, C., Najman, L., & LeCun, Y. (2013). Learning hierarchical features for scene labeling. *IEEE Transactions on Pattern Analysis and Machine Intelligence, 35*(8), 1915–1929.
10. Douze, M., et al. (2009). Evaluation of GIST descriptors for web-scale image search. In *Proceedings of the ACM international conference on image and video retrieval, Article No. 19, Greece.*
11. Ahmadi, M., Ulyanov, D., Semenov, S., Trofimov, M., & Giacinto, G. (2016). Novel feature extraction, selection and fusion for effective malware family classification. In *Proceedings of the 6th ACM conference on data and application security and privacy, Louisiana, USA.*
12. Han, K. S., Lim, J. H., Kang, B., & Im, E. G. (2015). Malware analysis using visualized images and entropy graphs. *International Journal of Information Security, 14*(1), 1–14.
13. Vinayakumar, R., Alazab, M., Soman, K., Poornachandran, P., & Venkatraman, S. (2019). Robust intelligent malware detection using deep learning. *IEEE Access, 7*, 46717–46738.
14. Kalchbrenner, N., Grefenstette, E., & Blunsom, P. (2014). A convolutional neural network for modelling sentences. In *Proceedings of the 52nd annual meeting of the Association for Computational Linguistics, Maryland, USA* (pp. 655–665).
15. Bhodia, N., Prajapati, P., Troia, F. D., & Stamp, M. (2015). Transfer learning for image-based malware classification. In *Proceedings of the 5th international conference on information systems security and privacy* (pp. 719–726).
16. Alex, T. *Malware-detection-using-machine-learning* [Online]. Available https://github.com/tuff96/Malware-detection-using-Machine-Learning

17. Su, J., Danilo Vasconcellos, V., Prasad, S., Daniele, S., Feng, Y., & Sakurai, K. (2018). Lightweight classification of IoT malware based on image recognition. In *2018 IEEE 42nd annual computer software and applications conference (COMPSAC), Tokyo* (pp. 664–669).
18. Le, Q., Boydell, O., Mac Namee, B., & Scanlon, M. (2018). Deep learning at the shallow end: Malware classification for non-domain experts. *Digital Investigation, 26*(1), 5118–5126.
19. Cui, Z., et al. (2018). Detection of malicious code variants based on deep learning. *IEEE Transactions on Industrial Informatics, 14*(7), 3187–3196.
20. Lad, S. S., & Adamuthe, A. C. (2020). Malware classification with improved convolutional neural network model. *International Journal of Computer Network and Information Security, 12*(6), 30–43. https://doi.org/10.5815/ijcnis.2020.06.03
21. Tareen, S. A. K., & Saleem, Z. (2018). A comparative analysis of SIFT, SURF, KAZE, AKAZE, ORB, and BRISK. In *International conference on computing, mathematics and engineering technologies (iCoMET 2018), Sukkur, Pakistan.*
22. Gurucharan, M. K. (2020). *Basic CNN architecture: Explaining 5 layers of convolutional neural network* [Online]. Available https://www.upgrad.com/blog/basic-cnn-architecture/#:~:text=There%20are%20three%20types%20of,CNN%20architecture%20will%20be%20formed
23. Agarwal, B., et al. (2020). *Deep learning techniques for biomedical and health informatics.* Academic.
24. Smeda, K. (2019). *Understand the architecture of CNN.* Towards Data Science [Online]. Available https://towardsdatascience.com/understand-the-architecture-of-cnn-90a25e244c7
25. Albelwi, S., & Mahmood, A. (2017). A framework for designing the architectures of deep convolutional neural networks. *Entropy, 19*(6), Article 242.
26. El-Baz, A. S., & Suri, J. (2021). *Neural engineering techniques for autism spectrum disorder.* Academic.
27. US Senate. (2022). *America's data held hostage: Case studies in Ransomware attacks on American companies.* Staff Report [Online]. Available https://www.hsgac.senate.gov/imo/media/doc/Americas%20Data%20Held%20Hostage.pdf
28. MachineLearningMastery. (2021). *How to develop a CNN from scratch.* Retrieved from https://machinelearningmastery.com/how-to-develop-a-generative-adversarial-network-for-a-1-dimensional-function-from-scratch-in-keras/
29. Rieck, K., Trinius, P., Willems, C., & Holz, T. (2011). Automatic analysis of malware behavior using machine learning. *Journal of Computer Security (JCS), 19*(4), 639–668.
30. Madhuanand, L., Nex, F., Yang, M., Paparoditis, N., Mallet, C., Lafarge, F., et al. (2020). Deep learning for monocular depth estimation from UAV images. *ISPRS Annals of the Photogrammetry, Remote Sensing and Spatial Information Sciences, 2*, 451–458.
31. Moussas, V., & Andreatos, A. (2021). Malware detection based on code visualization and two-level classification. *Information, 12*(3), 118.

Chapter 3
Malware Anomaly Detection Using Local Outlier Factor Technique

Abstract Malware anomaly detection is a major research area as new variants of malware continue to wreak havoc on business organizations. In this study, we propose a new technique based on the Local outlier factor algorithm to detect anomalous malware behavior. We empirically validate the performance and effectiveness of our technique on real-world datasets. This is an efficient technique for malware detection as the model trained for this purpose is based on unsupervised learning. The model trains on the anomalies, that is, the unusual behavior in a process, making it significantly effective.

Keywords Local outlier factor · Outlier detection · Anomaly detection · Malware detection · Malware dataset · Dataset validation

3.1 Introduction

The world is advancing at a rapid pace. State-of-the-art technologies are incorporated into every single domain involving the use of technology. The field of science and technology owes its vast expansion and advancement to the use of communications and information systems. With the applications of information technology expanding significantly over the past few decades, the number of users relying on the applications of communications and information technology has also increased significantly, evidently causing various concerns in terms of security and privacy.

Where it is evident that the IT domain has exponentially grown over the past few decades, the potential security threats have also increased. Cybersecurity is adopting new and innovative methods to counter the invasive approaches adopted by cybercriminals. The application and implementation of these advancements can be seen as the success in the domain of digital forensics and surveillance, malware and botnet, intrusion detection and prevention systems, etc. [1]. Cybercriminals all around the world are constantly deriving new methods and ways to breach the security and integrity of highly sophisticated systems as a challenge.

© The Author(s), under exclusive license to Springer Nature Switzerland AG 2022
M. Omar, *Machine Learning for Cybersecurity*, SpringerBriefs in Computer Science, https://doi.org/10.1007/978-3-031-15893-3_3

To address these issues, various intrusion detection and prevention systems are developed using the latest technologies like machine learning, artificial intelligence, federated learning, cryptography, etc. Artificial intelligence and machine learning have played a significant role in decision-making support, human behavior analysis, recommendations, process control, etc. One of the many applications of machine learning is in malware detection.

Employing machine learning for malware detection has many different approaches depending upon the threat vector and recognition algorithm for potential threats [2]. Machine learning techniques provide the ability to resolve complex problems and issues owing to the adaptation to changing circumstances. The applications of machine learning and artificial intelligence in the domain of cybersecurity address the problems such as phishing detection, network and application intrusion detection and prevention systems, cryptography, spam detection, denial-of-service prevention and DDoS prevention, etc. [3].

3.1.1 Malware Detection

Malware is one of the most persistent and significant security threats that damages computer systems resulting from a lack of cybersecurity practices. Malware detection is a process of the detection of malicious files or programs on the host system that can cause potential damage to the components of the system [4]. Malware development has come a long way progressing into the development of advanced persistent threat (APT) attacks, dynamic link library (DLL) attacks, etc. Such attacks are significantly difficult to detect and even further difficult to mitigate [5]. Malware is divided into different categories based on various factors, which include attack vectors, mode of delivery, lifecycle, threat vector, etc. Some common types of malware which are used to gain access and cause damage to systems by cybercriminals are Trojan horses, rootkits, botnets, worms, denial-of-services, etc. As a result of thousands of different variations of malware being developed and launched daily, security professionals develop various tools and software such as CWSandbox, ANUBIS, Norman Sandbox, etc. to detect and mitigate these threats [4].

3.1.2 Intrusion Detection Systems

As previously mentioned, cybercriminals devise thousands of new methods to exploit vulnerabilities in software and on the hardware level as well. To determine the threat and attack vector, security professionals develop different tools to detect and mitigate such attacks. Intrusion detection systems are software systems that detect the occurrence of malicious activity on information systems or network levels. These intrusion detection systems are explicitly developed to maintain the integrity, confidentiality, and availability of the data and information stored on computer systems [6]. Intrusion detection systems limit unauthorized access and escalated privileges for the attackers by employing different techniques. Due to its paramount importance, IDS have achieved a significant importance in every organization. The following reasons justify the use and implementation of IDS in the current era of technology:

- Prevention of problem behaviors by increased risk and threat discovery.
- Detection of exploits and attacks not mitigated by standard security controls.
- Detection and mitigation of "doorknob rattling" and network attacks.
- Documentation and report of existing threats and vulnerabilities.
- Quality and security control for administration and security design implementations.
- Improvised diagnosis and analysis of the existing and potential future threats along with recovery and vulnerability analysis [6].

Intrusion detection systems are of different types depending on the usage and requirements of users such as interval-based, real-time, network-based, host-based, application-based, etc. Each intrusion detection system implements a different strategy based on the protection criteria and vulnerabilities covering different aspects of the attack vectors. This study highlights the use and implementation of a network-based intrusion detection system.

3.1.3 Network-Based Intrusion Detection System

Network-based IDS are one of the most used intrusion detection systems employed by companies commercially. Network-based IDS monitor network traffic and inspect packets to detect unusual behavior and analyze different attacks before they reach the system through the network. One network-based IDS can listen into network segments of multiple hosts, thus providing real-time threat analysis to the entire network (Fig. 3.1).

In a more common approach, NIDS comprise two network interfaces; one interface is employed in promiscuous mode to listen in on the network information transfer, while the other is used for reporting and control. Feeding traffic as an input to NIDS has many different approaches; with the development of switching techniques, port-mirroring techniques supply traffic to NIDS. For example, Cisco employs the method of switched port analyzer for this purpose. Depending upon the functionality and usage requirements, NIDS can either be anomaly-based or signature-based [8].

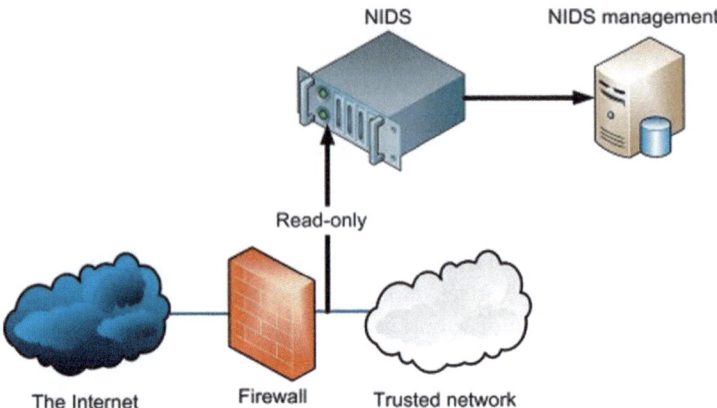

Fig. 3.1 Network-based IDS architecture [7]

3.1.4   Advantages of Network-Based Intrusion Detection System

- Large networks extending over multiple nodes can be easily monitored by a few IDS.
- The incorporation of network-based IDS into an already established network requires minimal efforts with the assurance of significantly little impact on the network. NIDS are passive systems that when implemented only listen to the network traffic without causing any hindrance to the normal functionality and operations of the network.
- The implementation and deployment of NIDS can be made significantly secure and protected against different ranges of attack. The implementation of NIDS can even be made invisible to the attacker.

As previously mentioned, the detection and mitigation of different attacks having a vast range of variables can be achieved by the implementation and employment of advanced machine learning techniques. The focus of this study is to understand and implement a machine learning approach known as the "local outlier factor" to mitigate anomaly-based attacks.

3.2   Related Works

The anomaly detection approach has been developed by various researchers in the recent past. One of the approaches employed by Kim et al. [9] uses the local outlier factor to determine the unfamiliar intrusions in an endpoint environment. Kim et al. [9] employ the use of two models for the detection of unusual behavior that depicts deviation from normal circumstances. Anomaly detection is achieved by the analysis of the event log. This generates the anomaly scores. Based on the frequency of the event occurrence, the attack types are identified. Employing the use of local outlier factor, and Autoencoder, the model is designed to detect anomalies efficiently. Based on different attack scenarios presented by the attacker, the analysis of the attack profile is used for the detection of anomalies. The research conducted by Kim et al. [9] indicates that various studies have been conducted for the detection of supervised learning-based behaviors of attacks by employing the use of labelled data such as denial-of-service (DOS), user-to-root (U2R) attack, remote probing attacks, and remote-to-local (R2L) attack. As compared to the studies highlighted above, the research carried out the implementation of the model that determines the deviation of behavior from normal behavior under ordinary circumstances. The proposed model operates using LOF and Autoencoder for the calculation of anomaly score which represents the data points between data gathered from the existing logs and newly generated logs (Fig. 3.2).

Kotu and Deshpande [10] proposed various methods of anomaly detection in datasets. The researchers explain in great detail the different methods to detect anomalies that occur in datasets. The two methods highlighted among the different methods proposed are outliers detection using statistical methods and outlier detection using data mining. By employing a statistical distribution model, the data points that do not align with the model are identified. Outliers are detected based on the location of the data points in the standard deviation curve. By applying the data mining method for outlier detection, various parameters and variables highlight the outliers in data. These variables are distance, density, distribution, clustering, and classification techniques. Each of the variables and their working are explained in detail by Kotu and Deshpande [10] in the research carried out.

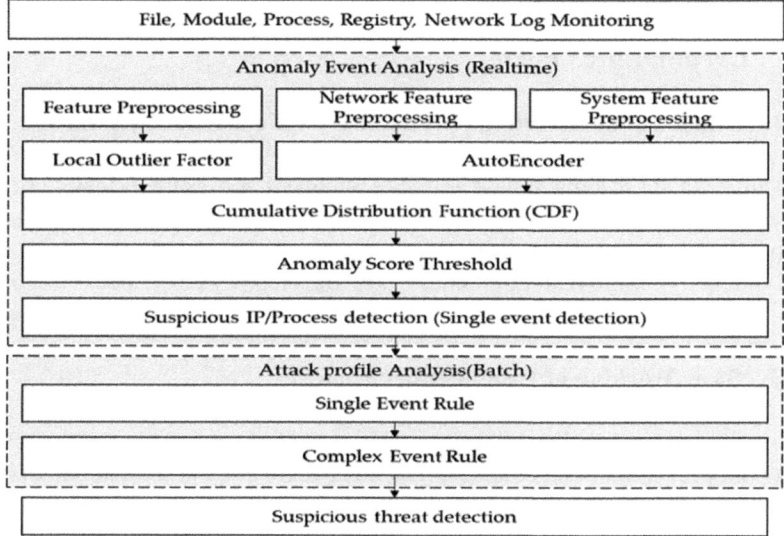

Fig. 3.2 Proposed model by Kim et al. [9]

The research and study conducted by Tang et al. [11] emphasize the detection of malware while the exploit or an attack is being carried out. This strategy acts as an early-threat detector for the improvement in the accuracy of the detection of payloads while they are being delivered. The fundamental intuition behind the operation of an anomaly-based malware detection methodology is to analyze the alteration in the normal execution of a program by a peculiar non-native script or a code block. If these perturbations are identified and analyzed, they can form the fundamentals of malware detection. The study conducted by Tang et al. [11] uses Internet Explorer 8 and Adobe PDF Reader 9 for the determination of attack vectors as these programs are more commonly exploited by the attackers.

3.3   Proposed Methodology

The introduction of advanced machine learning in the domain of cybersecurity has significantly decreased the risks associated with the use of technology. As previously mentioned, various machine learning techniques are applied to counter the attacks by cybercriminals. The use of both supervised and unsupervised learning has opened new dimensions for the protection of critical data and information. Various machine learning techniques such as decision trees, similarity hashing (locality-sensitive hashing), incoming stream clustering, behavioral modeling, etc., are employed for the security of information [12]. One of the machine learning methods used for the detection of malware is the local outlier factoring technique which is the fundamental method employed and implemented in this research.

3.3.1   Local Outlier Factor

3.3.1.1   A Brief Understanding of the Local Outlier Factor Approach

Local outlier factor is a machine learning technique that uses the principle of unsupervised learning for outlier detection. The outliers in a dataset are data points that are represented as anomaly scores. This is achieved by calculating the deviation of the local density for the data point concerning the closest data points.

3.3.1.2   Basic Working of Local Outlier Factor

The determination of the local density is achieved by the estimation of the distance between different data points that are close to each other (k-nearest neighbors). The density for each data point can be measured by the comparison of the data points having similar densities with low-density data points in the vicinity. The low-density data points are considered outliers [13].

Fig. 3.3 Relative
densities in LOF [14]

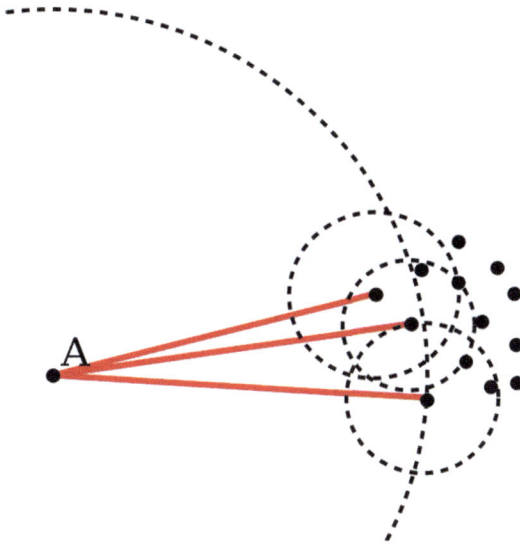

reachability-distance$_k$(A,B)=max{k-distance(B), d(A,B)}

Fig. 3.4 Reachability distance between two points and k-distance of the point [13]

Fig. 3.5 Local reachability
distance formula [13]

$$\mathrm{lrd}_k(A):=1/\left(\frac{\sum_{B\in N_k(A)}\text{reachability-distance}_k(A, B)}{|N_k(A)|}\right)$$

The following image shows the basic classification of the local outlier factor
(Fig. 3.3).

Initially, the distances between the two points are k-distances. These data are
relatively close to each other; hence, the distances between each point are for the
determination of their k-nearest neighbors. The second nearest point is the closest
neighbor to the initial point. This is used to measure the reachability distance which
is the maximum distance between two different points and the k-distance of that
specific point (Fig. 3.4).

3.3.1.3 Local Reachability Distance

The local reachability distance of the k-nearest points close to a specific point is
measured by the calculation of the inverse of the sum of all the k-nearest points for
the reachability distance. If the points are closer to each other, the distance is low,
and hence the density is increased (Fig. 3.5).

Fig. 3.6 Local outlier factor formula [13]

$$\text{LOF}_k(A) := \frac{\sum_{B \in N_k(A)} \frac{\text{lrd}_k(B)}{\text{lrd}_k(A)}}{|N_k(A)|} = \frac{\sum_{B \in N_k(A)} \text{lrd}_k(B)}{|N_k(A)| \cdot \text{lrd}_k(A)}$$

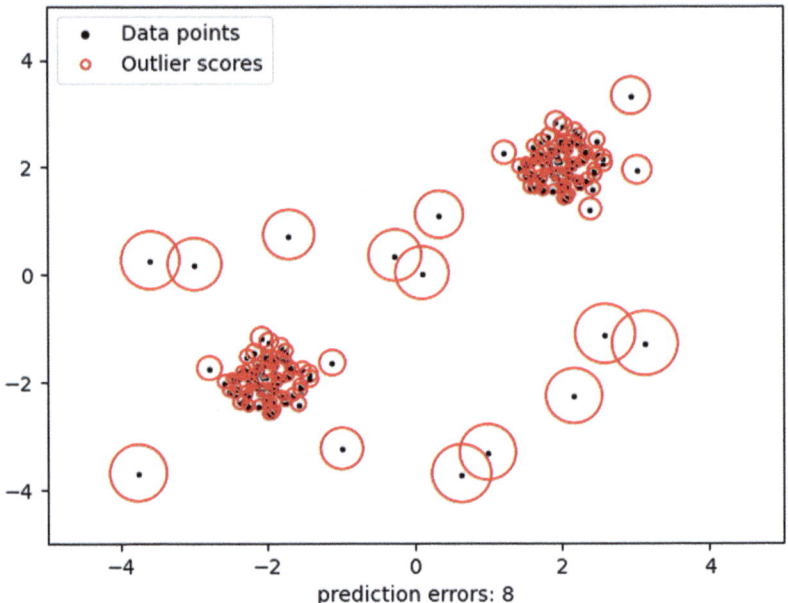

Fig. 3.7 Implementation result of local outlier factor on a dataset [15]

The local outlier factor (LOF) is calculated by the ratio of average local reachability distances of neighbors and the local reachability distance of that point (Figs. 3.6 and 3.7).

3.4 Results and Discussion

We conducted our experiments on the much popular NSL-KDD dataset. NSL-KDD is a new version dataset of the KDD'99 dataset (see details in Table 3.1). This is an effective benchmark dataset to help researchers compare different intrusion detection methods. In the following figure, we present the results of our LOF technique and its performance on the NSL-KDD dataset (Fig. 3.8).

In the following figure, we illustrate the prediction performance of our LOF technique on the NSL-KDD dataset (Fig. 3.9).

Python Code
Below is a snippet of the python code which we trilized in running our experiemnts on the KDD dataset using our LOF-based technique.

Table 3.1 The NSL-KDD's feature description

Feature name	Description
Duration	Length (number of seconds) of the connection
Protocol type	Type of the protocol (e.g., tcp, udp, etc.)
src bytes	Number of data bytes from source to destination
dst bytes	Number of data bytes from destination to source
srv count	Number of connections to the same service as the current connection in the past 2 s
dst host same src port rate	Number of connections that were to the same source port

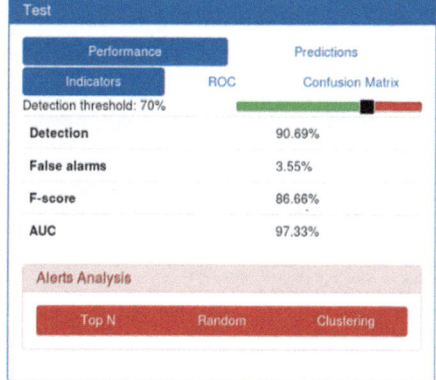

Fig. 3.8 Performance of LOF technique on the NSL-KDD dataset

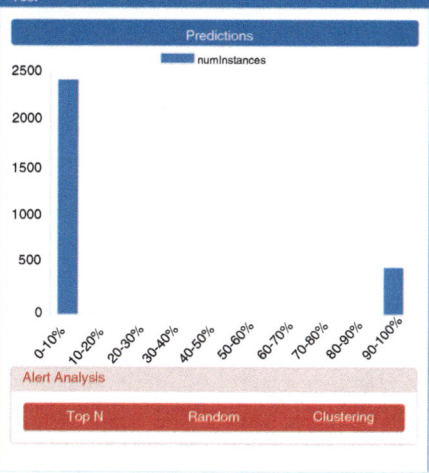

Fig. 3.9 Prediction performance by LOF technique on the NSL-KDD dataset

```
import numpy as np
from sklearn.metrics import roc_auc_score, roc_curve

warnings.filterwarnings("ignore",    message="numpy.dtype    size
changed")
with warnings.catch_warnings():
 warnings.filterwarnings("ignore",category=DeprecationWarning)
warnings.filterwarnings("ignore",    message="Using    a    non-tuple
sequence")

import copy
import matplotlib.pyplot as plt
from mlxtend.evaluate import confusion_matrix
from modules import utils, dimension_reduction as dim_red, evalu-
ation as eval, clustering as cluster
import sys

try:
 path = sys.argv[1]
except IndexError:
 is_product = False
else:
 is_product = True

DIMENSION = 30
SEPARATOR = "==============================\n"

# 0. Data loading
if is_product:
 train, ytrain = utils.load_train_data(path, is_product)
else:
  train, ytrain = utils.load_train_data('./data_in/satan_normal.
csv', is_product)

# 1. Dimension Reduction
T = DIMENSION
n = train.shape[0]
projected = dim_red.pca(train, T, is_product)

# 3. Clustering
predict = cluster.LOF_score(projected)
train["rate"] = predict
train["label"] = ytrain

# 4. Evaluation
```

```
if is_product:
 for i in train["rate"]:
 print(i)
else:
 fpr, tpr, threshold = roc_curve(ytrain, train["rate"])
 t = np.arange(0., 5., 0.001)
 utils.plot(1, 1, fpr, tpr, 'b', t, t, 'r')
 print("AUC score : ", roc_auc_score(ytrain, train["rate"]))
 print("finish")
```

3.5 Conclusion

The research conducted for this study highlights in detail the role of various emerging technologies in the domain of cybersecurity. As the reliability of every single user of the internet has increased significantly in the applications of modern technology, the risks associated with the extensive use of these technologies also arise. While it was evident that technology has evolved significantly over the last few decades, cyberattacks have also increased exponentially. These cyber-attacks compromise the confidentiality, integrity, and availability of the data while damaging critical infrastructure. To counter such invasions of privacy, security researchers and professionals are constantly devising new methods and techniques to overcome security challenges. Advanced machine learning techniques, such as decision trees, local outlier factor, k-nearest neighbor, convolutional neural networks, etc., are used to counter these attacks. This paper addresses the problem of detection of malware using the local outlier factor. This is an efficient technique for malware detection as the model trained for this purpose is based on unsupervised learning. The model trains on the anomalies, that is, the unusual behavior in a process, making it significantly effective.

References

1. Anbar, M., Abdullah, N., & Manickam, S. (2019). *Advances in cyber security*. Springer.
2. Nayyar, S. (2021). *Council post: What machine learning can bring to cybersecurity*. Forbes. Retrieved 02 May 2022 from https://www.forbes.com/sites/forbestechcouncil/2021/10/01/what-machine-learning-can-bring-to-cybersecurity/?sh=72e575e01203
3. Ford, V., & Siraj, A. (2014). Applications of machine learning in cyber security. In *Proceedings of the 27th international conference on computer applications in industry and engineering* (Vol. 118). IEEE Xplore.
4. Katzenbeisser, S., Kinder, J., & Veith, H. (2011). Malware detection. *Encyclopedia of Cryptography and Security*, 752–755. https://doi.org/10.1007/978-1-4419-5906-5_838
5. Guezzaz, A., Benkirane, S., Azrour, M., & Khurram, S. (2021). A reliable network intrusion detection approach using decision tree with enhanced data quality. *Security and Communication Networks, 2021*, 1–8. https://doi.org/10.1155/2021/1230593

6. Bace, R., & Mell, P. (2001). Intrusion detection systems. *NIST, 51*. Retrieved 02 May 2022 from http://cs.uccs.edu/~cchow/pub/ids/NISTsp800-31.pdf

7. Conrad, E., Misenar, S., & Feldman, J. (2017). Domain 7. *Eleventh Hour CISSP®*, 145–183. https://doi.org/10.1016/b978-0-12-811248-9.00007-3

8. Introduction to Intrusion Detection Systems. (2003). 1–38. https://doi.org/10.1016/b978-193226669-6/50021-5

9. Kim, S., Hwang, C., & Lee, T. (2020). Anomaly based unknown intrusion detection in endpoint environments. *Electronics, 9*(6), 1022. https://doi.org/10.3390/electronics9061022

10. Kotu, V., & Deshpande, B. (2015). Anomaly detection. *Predictive Analytics and Data Mining*, 329–345. https://doi.org/10.1016/b978-0-12-801460-8.00011-2

11. Tang, A., Sethumadhavan, S., & Stolfo, S. (2014). Unsupervised anomaly-based malware detection using hardware features. *Research in Attacks, Intrusions and Defenses*, 109–129. https://doi.org/10.1007/978-3-319-11379-1_6

12. Kaspersky. (n.d.). Kaspersky.com. Retrieved 05 May 2022 from https://www.kaspersky.com/enterprise-security/wiki-section/products/machine-learning-in-cybersecurity.

13. GeeksforGeeks. (2020). *Local outlier factor—GeeksforGeeks*. GeeksforGeeks. Retrieved 3 May 2022 from https://www.geeksforgeeks.og/local-outlier-factor/

14. ajulsam. (n.d.). *Local outlier factor | Graph lab create user guide*. Ajulsam.gitbooks.io. Retrieved 03 May 2022 from https://ajulsam.gitbooks.io/graphlab-create-user-guide/content/anomaly_detection/local_outlier_factor.html

15. sckit-learn. (n.d.). *Outlier detection with Local Outlier Factor (LOF)*. scikit-learn. Retrieved 05 May 2022 from https://scikit-learn.org/stable/auto_examples/neighbors/plot_lof_outlier_detection.html#:~:text=The%20Local%20Outlier%20Factor%20(LOF,lower%20density%20than%20their%20neighbors

Printed by Printforce, the Netherlands